D1643478

TORRIDON

Torridon

Lea MacNally

SWAN·HILL
PRESS

Lea MacNally died suddenly on 14 February 1993, having completed the text of this book but without leaving a dedication to it. We think the dedication would have gone something like this:

'To Margaret, who quietly and unselfishly shared the ups and downs of our twenty-one years in Torridon, and to our grandchildren, Margaret Ann and Catriona who spent many happy holidays with us there.'

Lea and Michael MacNally

Copyright © 1993 by Executors of Lea MacNally

First published in the UK in 1993
by Swan Hill Press an imprint of Airlife Publishing Ltd

British Library Cataloguing in Publication Data
  A catalogue record for this book
  is available from the British Library

ISBN 185310 350 0

All rights reserved. No part of this book may be reproduced or transmitted in any form or by any means, electronic or mechanical including photocopying, recording or by any information storage and retrieval system, without permission from the Publisher in writing.

Printed in England by Livesey Ltd., Shrewsbury

Swan Hill Press

an imprint of Airlife Publishing Ltd
101 Longden Road, Shrewsbury SY3 9EB

Contents

Author's Introduction

Torridon is not, and never will be, cosily domesticated! If you do not enjoy living in surroundings of stark and sometimes hostile grandeur, then do not live in Torridon. If you cannot tolerate heavy rain, salt spray and high winds, then do not live in Torridon. If you do not relish having to drive sixty miles to a hospital, then do not live in Torridon. If you do not like living in an area where amenities, so called, are limited, then eschew Torridon. If you cannot tolerate living in an area where there are no supermarkets and where the village shop charges more for the basic necessities of life, simply to survive, then do not try to live in Torridon.

To live in Torridon you must enjoy, appreciate *and accept Torridon* for what it is. It is quite pointless to try to soften its hard edges. It is the hard edges which render Torridon memorable! You may justifiably feel that, in this book, I have given undue emphasis to enthusing over wild country and its sparse and specialised wildlife. But Torridon, you know, is a wild place; Nature so designed it and in an increasingly urbanised and artificial world that is what confers on it its uniqueness, its attraction. Most of all this book is an appreciation of its naturally wild places and the way of life implicit, nay, enforced there. There is an ages-old brooding serenity in Torridon which owes nothing, absolutely nothing, to the artefacts of the human race which indeed succeeds only in detracting from this serenity. The tranquil quality of an evening sunset as one looks down the long, mirror-like, glassy calm of Loch Torridon, fringed by the darkening, brooding, secretive hills, sky and loch alike tinted a lovely, soothing lemon yellow, can hardly be matched in all of Britain.

Human occupation of Torridon has always been grudgingly conceded, thinly scattered around its lower fringes, and this is so to this day. Even those hardy animals, the red deer, find their higher mountain fastnesses untenable when winter gales howl and then they too, seek the sea-loch fringes of these mountains. In this over-crowded, increasingly polluted, urbanised country, I feel that we should accept Torridon's intolerance of human occupation and all its implications. Let us be graceful in our acceptance, and grateful, also. The National Trust for Scotland has guarded Torridon well, against undesirable development. I have no doubt that they will continue to do so. Torridon is worthy of such care.

1. Introduction to Torridon

In the spring of 1969 my life changed dramatically. It was then that I got the opportunity to go to live and work at Torridon, a mountainous area of Wester Ross, a spectacularly scenic area of mountains, rivers and sea-lochs, one which was a Mecca to hillwalkers throughout Britain and Europe. W. H. Murray, mountaineer and author, wrote of the area, 'Glen Torridon, its loch and the mountains to either side, exhibit more mountain beauty than any other district of Scotland, including Skye'. This estate of Torridon had come into the ownership of the National Trust for Scotland in May 1967, while early in 1968 the Trust was presented with a further 2000 acres adjoining and immediately to the west of Torridon estate thus making a total of more than 16,000 acres of superlative scenery. On the N.W. boundary of Torridon is the shattered quartzite summit ridge of Beinn Eighe (3309 ft); on its southern boundary there rises the enormous bulk of Liathach (3456 ft) dominating the glen below, with its seven tops linked by five miles of narrow, jagged ridges. To the west is Beinn Alligin (3232 ft). In addition to their scenic grandeur these mountains are of great interest to geologists. Liathach and Beinn Eighe are composed of red Torridonian sandstone estimated to be 750 million years old while their summit ridges are partly of grey-white quartzite, a mere juvenile at 600 million years old. This younger rock, the quartzite, contains pipe-rock, some of the earliest forms of fossil life.

In 1969 I was about to leave Culachy, that deer-forest above Fort Augustus and Loch Ness which had been home to Margaret and me, and our two sons, all of our nineteen years of married life. This estate had changed hands three times in our nineteen years there. Two owners we had liked, trusted and respected – the third owner justified the wry Highland adage that 'the steps of the Big House are *always* slippery'. We had been very happy at Culachy and I had got to know and love the hills intimately, with the deer, eagles and other fascinating wildlife present there. I had learnt a lot there, not least about human nature; now the die was cast, I had finally slipped on the steps.

Word had got around that I was leaving and on that spring day there arrived at the house a Colonel James Stewart from the National Trust for Scotland, empowered to approach me, on behalf of the Trust, to find out if I would consider employment with NTS on their recently acquired mountainous property of Torridon. There existed on Torridon a resident population of red deer; they had heard that I was keenly interested in the proper management of hill red deer. Would I undertake this at Torridon?

It was a complete surprise and an intriguing proposition. I knew comparatively little about Torridon then, nor of its red deer population. However it was hill country and, with the deer and other wildlife implicit in this, it was my kind of sphere. I agreed! Little did I know that the mountains of Torridon bore absolutely no resemblance to the rounded ridges of the Monadhliaths with which I was so familiar.

The next step was an arranged formal interview down in Edinburgh. The H.Q. staff of the National Trust for Scotland was refreshingly small in 1969. My interview completed, we foregathered for an anything but formal pub lunch in a bar conveniently near to 5 Charlotte Square. Colonel Stewart, Sir 'Jamie' Stormonth Darling, Robin Prentice, Donald Erskine, Finlay McQuarrie and Phil Sked, (all NTS stalwarts of the era) were there, and all were friendly and approachable. I decided I had made the correct decision. Upon being asked how long I would guarantee to stay at Torridon I replied 'at least five years'. I was to stay there for twenty-one years.

We moved to the Mains, a house at the head of Loch Torridon, in July 1969. 'We' included my tame deer, suitably crated, for which, with full approval and help from the Trust, I had had a fenced enclosure made behind the house.

That very first night, tired as we were, we could not get to sleep for the rasping, repetitive call of a corncrake, a sound I had not heard at Fort Augustus since the mid '50s when, there, they had simply ceased to breed in the fields. This was about the time when the making of silage replaced the traditional haymaking and the silage crop, being cut much earlier in the year than the hay had been, spelt disaster for the nesting corncrake. Yet here they were at Torridon, nesting still; this was a welcome bonus right away. Nor, I was to find, was this the only interesting bird on our doorstep; on a wide boggy flat not 200 yards from the Mains, at the head of the flat estuary of Loch Torridon, there nested both redshank and greenshank, both species vociferous in their protective anxiety of their nearly-fledged young. I was well aware that redshanks nest on low-level marshy ground, often close to a loch, but I had never before experienced greenshank at sea level. My nesting greenshanks, at Fort Augustus had been at an altitude of around 2000 ft, and remote from any habitation.

On the shingle shores of Loch Torridon there were ringed plover nesting and out on the loch were rafts of eider duck while whistling oystercatchers flashed by to land on the mussel-festooned rocks. We had only been in residence for days when I sighted three golden eagles from the house windows, soaring in flight high over the mountains opposite. In later years I was to visit their eyrie. Over the years I was to see from the house windows, red deer, fox, otter, pine marten, stoat and weasel. I had never had that experience at Culachy, remote as our house was on the edge of the hill. There proved to be a wide variety of both birds and animals at Torridon though the actual numbers of any one species were not high. That the terrain went, quite literally, from sea level up to nearly 3500 ft, and that this included salt water lochs, fresh water lochs, moorland, mountains and a little woodland, undoubtedly contributed to this variety. It was also sparsely inhabited, for it was, again literally, harsh, sterile, uncompromisingly rocky country, incapable of supporting a high human population. The fact that the hills were composed so largely of rock, and of basically infertile rock at that, meant that it could not support a high number of grazing animals, such as red deer, sheep, mountain hares or even voles. All were present but in relatively small numbers. Grouse were not plentiful, but that beautiful white grouse of the high ground, the ptarmigan, seemed to thrive on the stony screes of the rocky Torridon hills. Since prey species were not present in high numbers their predators were also relatively few in number and they concentrated on the areas where prey was more available. The fox, for instance, concentrated more around the coastal belt rather than the mountains inland. There, at least in time of need, it could scavenge the shore.

The hill lochs held both species of diver which habitually breed in Scotland, the supremely elegant black-throat, and the smaller and more numerous red-throat. Both were a joy to watch and to photograph. One pair of red-throats nested within a foot of the water's edge of a small dubh lochan. I had some magnificent views of both the adult birds with their dove-grey heads, striped black and white down the back of the neck and wine-red on the lower throat. By mid July this pair had one sooty black morsel of a chick swimming behind them on the lochan for only one of their two eggs hatched. The family was still there in August, the well-grown 'chick' by then swimming and diving as expertly as the parents. On the actual sea loch I was to learn that the biggest of the diver family, the great northern diver, turned up every winter. A massive bird, to me it lacked the grace of both red-throat and black-throat though, of course, like both of these, it was a superlative swimmer and diver.

I could hardly wait to get settled in so that I could get out to try the hill. It could hardly have been more different from my native

Monadhliaths. I had never seen so much naked rock, as I saw at Torridon, with the hills rising, often vertically, from sea level to 3000 ft. I became used to this lung-testing, sharp ascent without any approach work over flats and foothills and indeed I came to appreciate it. A mountaineer's paradise and, given the right weather, one of incomparable scenic grandeur, the more spectacular because one so seldom got the right weather.

On one of my first ventures into the unknown I went up Coire Dhubh, that gulf which separates Liathach from Beinn Eighe, to walk around the western buttress of Sail Mhor into the impressive glacial coire of Coire Mhic Fhearachair, accompanied by Michael, my younger son. Lea, our eldest son, had stayed 'back home' where he was employed in learning deer management. The path up into Coire Dhubh was steep, rugged and rocky; I was to find that unless one was willing to accept this steep, rugged and rocky terrain one did not go to walk on Torridon. Branching off the main path an even rougher track led eastwards, at times over rocks, at others through eroded patches of boggy peat, to Coire Mhic Fhearachair. The last stretch of the approach was steep and leg-testing; we came over a rocky rim, boulder-scattered, to see a handsome waterfall spill, white and sparkling, over a natural rock sill. Up over the rock sill and the huge bowl-shaped mountain-girdled coire, with grey screes spilling down to a jet-black loch, unfolded before us. It was a truly primaeval scene, elemental, stark, basic, culminating in shattered rock, black water and, at the loch's head, the forbiddingly steep precipices of the Triple Buttress, flanked by more aprons of rock scree. Yet in this remote wild coire there could be seen evidence of humans not, thank goodness in the shape of litter, but rather more tragic evidence.

As we picked our way around the rocks of the loch's shore, seeing the flowers of cow-wheat and dwarf cornel growing among them, I suddenly saw, ahead, high on the scree, a huge rubber-tyred wheel. Puzzled, we stopped and spied the huge expanse of scree below and N.W. of the Triple Buttress cliff. We began to pick up pieces of alloy fuselage; a huge aircraft engine and the vertical blade of a propeller reared among the rocks. Quite obviously there had been an aircraft crash here and, absolutely unaware of when it had happened, it was with some apprehension that we went to investigate. Drawing nearer, I saw that there was thick rust on the framework of the huge wheel. From then on curiosity replaced apprehension and we separated, poking among the scattered rocks of the scree. Fragments, some small, some huge, were everywhere. On some of the smaller fragments of alloy toothmarks showed, where red deer, with their penchant for chewing hard objects such as bone or antler, had tried, as a variation, an aeroplane. Looking down on the loch an entire engine could be seen lying in the green-tinted shallows.

It had been a huge plane, a Lancaster bomber, and I later heard its story. On 13 March 1951, it has been on a navigation exercise flying from Kinloss, towards Rockall and the Faroes. On the last leg of the exercise returning to Kinloss at an estimated time of about 2 am and flying totally on instruments it had flown into the sheer rock top of the Triple Buttress at 2850 ft. Two days of unsuccessful searching ensued before a laddie in the Torridon area, (Murdo MacDonald, I believe) reported seeing an inexplicable red glow in the night sky over Beinn Eighe. Subsequently on 17 March, a searching aircraft located the wreckage. There was very heavy snow in that year and over the next few days repeated attempts to reach the spot, by a rescue team from RAF Kinloss failed completely. The use of helicopters for mountain rescue work was not yet an established practice. Offers of help from skilled snow-climbers of the Scottish Mountaineering Club were apparently declined by the RAF. Security, perhaps?

Eventually two Royal Marine commandos managed to reach the wreckage on the top and made the grim discovery of a body near the cairn on Sail Mhor. They also established that none of the eight crew could possibly have survived. By the end of that March three more bodies were found in a section of the fuselage overhanging the Triple Buttress. It was, I was told, the end of August before the last body was eventually recovered. I was also told that it was because of this tragic mountain accident in such remote country, even though only sixty miles from Inverness, that the very intensive upgrading of the RAF mountain rescue facilities took place. It was the May of 1974 before our Torridon mountain rescue team first had experience of the tremendous aid which the helicopters of the RAF mountain rescue service could give. Nowadays the use of mountain rescue helicopters is taken for granted and many lives have been saved by their dedicated crews.

Our curiosity assuaged, we headed back by the exceedingly rough and rock-scattered North face of Liathach. We saw around one hundred red deer that day, some in places and heights that led me to surmise that the deer of Torridon had a strain of chamois in their lineage! Indeed I came to find that stalking red deer on Torridon was more akin to stalking chamois than any stalking I had previously experienced. On the way back I photographed a hind and her calf who were engrossed in watching a file of hill walkers, clad in bright oranges, reds and blues, passing along a glen path far below. One realises that the modern penchant for such garish colours is because they are glaringly obvious to a rescue party in the event of mishap, but one also realises why hill walkers rarely see much of the wildlife around them, clad in this fashion.

Bird life on the hill that day was scarce but very interesting. We saw no less than three eagles, typical of this time of year, a pair with their recently-fledged young one now under parental instruction. One of the adults gave us a fine display when, head on into a gale force wind, it hung suspended in the sky like a huge dark-winged kestrel without a movement of its stiffly-held wings or widespread tail feathers. Grouse droppings and some feathers we found but saw no grouse. Greenshank we heard, anxiously piping, but utterly failed to see. We watched four adult black-throated divers, a pair on one good sized lochan and a single bird on two other lochans, but their young must have been still on their nesting lochs. Young ptarmigan we did see. When we were crossing a scree of grey, splintered, moss and fern-grown rocks, a hen ptarmigan materialised magically out of nowhere and fluttered enticingly ahead of us, decoying us away from where, perched on a rock, stood a well-grown young one. It took wing even as we speculated as to whether it was the only survivor of a brood. As if in answer, five fragments of 'grey rock' became animated and flew from us, with the lovely flutter of pure white wings so typical of ptarmigan in flight.

A last steep ascent that day took us up and over the summit ridge of Liathach. Far, far below us were the matchbox-sized white houses of the village of Fasag, with Loch Torridon almost lapping their thresholds. A potentially breathtaking panorama was denied to us by the greyness of rain sheeting in from the sea. That, too, I was to get accustomed to, in the years to come, at Torridon.

While Liathach is unquestionably the grimmest of the Torridon giants, with three of its tops over 3000 ft and its exposed, airy, narrow Pinnacles ridge, strenuous and nerve-testing for most mountaineers, its other two mountains, Beinn Alligin (3232 ft) and Ben Dearg (which only just fails to reach 3000 ft) also require care and respect. These last two only dwindle in stature in relation to Liathach, so near and so dominant. The first day I ventured on the traverse of the ridge of Liathach I was again accompanied by Michael, still on his school holidays. On the top it was blowing a gale, though we had left only a moderate wind in the glen below. There was therefore no question of a first attempt on the exposed Pinnacles section for this is tricky enough even on a still day with its knife-edged series of jagged rock scrambles dropping away sheer on both sides. Disappointed, we skirted this section by the exceedingly narrow track which runs, ribbon-like, on the south side of the Pinnacles ridge. This track requires an impeccable head for heights, (vertigo sufferers, please note) and is innocuous only in comparison with the actual ridge. It is bordered to the south by a sheer drop and from its edge one looks *straight* down to the road, far, far below, in Glen Torridon, with its ant-like vehicles creeping along its narrow track. We enjoyed that high walk, though denied the ultimate thrill of the ridge.

We saw red deer, often as only red dots, way down in the northern coires of Liathach; a pair of eagles, patrolling ravens and, yet again, ptarmigan we also saw. One could see all of these in the Monadhliaths and also the pine marten which was just becoming evident there in 1969. The pine marten is a member of the weasel family, albeit a very beautiful one, and made its growing presence most evident by its habit, when natural prey was scarce in winter, of occasionally seeking food in the nearest henhouse (particularly if the henhouse was near to woodland). Having entered a henhouse at night the fluttering, cackling, blind panic of the occupants usually inflamed the killing instinct of the pine marten so that it was not satisfied until all these noisy birds were silenced. One could hardly blame the poultry owner for trying, legally or not, to exact vengeance and, since the pine marten is not trap-wary, this usually resulted in a trapped marten. While one regrets the killing of such a beautiful vital animal as the pine marten, with its lovely dark glossy chocolate-brown coat enhanced by a frontlet of vivid colour, varying from a deep yellow to an even deeper orange in individuals, one can also appreciate the 'eye-for-an-eye' feelings of the luckless poultry owner. The real remedy of course is to see that your hens are *securely* shut in each and every night. Human nature and indolence being what it is, and West Highland henhouses being as they are, this is, alas, more a counsel of perfection than one of probability.

On the coast, Torridon scores in its sea birds; ringed plover, dunlins, turnstone are on its sea loch shores, while rafts of eider ducks float out on the loch, their delightful drowsy crooning reminiscent of the sleepy *coo-rooing* of the cornfield woodpigeons. Gulls are legion, as are oystercatchers; in winter, curlew solemnly stalk the fields by my house, probing their damp surface with long curved bill. Mallard are also plentiful then and while I enjoy a roast mallard duck as much as anyone I get still more pleasure out of seeing them dabbling about on the fields while their exuberant quacking is welcome to my ears. Wigeon arrive in select numbers to the estuary, occasionally a colourful shelduck will appear, while heron are commonplace along the tideline. Otters, scarce in the Monadhliaths, were much more likely to be seen in Torridon, in sea loch, river, or very occasionally, on the hill. This being so it is the more regrettable that I heard, some time after we came to Torridon, that a family party of mother and two cubs had been mercilessly and wantonly shot as they played on the ice of a frozen pool in Glen Torridon. This was in the winter of 1968 when, to many folk, otters, as competitors for fish, were fair game. The situation, I am delighted to be able to write, is very different now; in the 1990s anyone who molests an otter is a virtual outcast.

Roe deer in Torridon are scarce and localised, logically so, since suitable woodland is also scarce. Japanese sika deer were absent completely in 1969 but had begun to penetrate the area when I retired in 1991. Badgers we had not got, though an occasional vagrant appeared from Strathcarron; Torridon was altogether too sterile and rock-bound to be a badger habitat. Fox was a different case altogether; while the fox population was much less than in the Monadhliaths the main crop of the Torridon crofts on NTS property was of lambs. Inevitably then this meant conflict between crofter and fox – inevitably too it meant that I, as NTS representative, was appealed to for help whenever a fox was 'convicted' of lamb-killing. Of one such punitive expedition, to the primaeval coire on Beinn Alligin known as Toll a'Mhadaidhe, I retain vivid memories.

The 'phone rang one evening as, pleasantly weary after a long day on the hill, I relaxed by the fire. The gist of the message was that the Alligin crofters had been losing lambs to a dismaying degree and had gone searching for a fox den. This they had discovered at last. Could I come out to it and wait out all night, with my rifle, in the hope that I might be able to shoot the absent vixen as she returned to her den as dusk fell? I could! I got my rifle, some food, and my sleeping bag and set forth.

The den, I discovered, was high up in the coire of Toll a'Mhadaidhe. It was in the huge jumble of gargantuan riven rocks which, aeons ago, when the coire was yet ice-bound, had raged down, in a cataclysmic avalanche of monstrous boulders, from Ben Alligin. This had left an enormous vertical gash cleft into Sgurr Mhor, Ben Alligin's highest peak, instantly recognisable in its Gaelic name of Eag Dhubh ('Black Notch'). Toll a' Mhadaidhe is the Gaelic for the 'Hole of the Wolf', or 'Den of the Wolf', and one can, to this day, very easily visualise its occupation by the wolf. There is a local tradition that the last wolf in this area of Wester Ross was in fact killed there. Gaelic place names are almost invariably descriptive and I, personally, put full credence in this claim. Nowadays this chaotic maze of ages-old, sharp-angled rocks, covering all the floor of the huge coire, harbours only the hill fox and a quite unexpected number of trees. A come-down in stature, from its days as a wolf sanctuary, yet it shelters the fox just as efficiently as ever it did the larger wolf. There are rocks as big as, and infinitely bigger than, motorcars, and gulfs, crevices and long heather-grown crannies, deep and part-hidden, between the rocks, enough to hide an army of men, never mind a fox's litter.

The discovered fox den had its fox-cub-trampled earth floor littered with prey and was highly redolent with the acrid scent of fox and of decaying prey remains. It was a good twelve feet down, sheltered and hidden between the sharp-edged fangs and spires of gigantic rock. It was ideal from the point of view of the fox cubs for they could play, feed, fight and exercise, unseen by man or, more importantly, by the pair of golden eagles which frequented the same coire. Most of the prey remains visible were, regrettably, of lamb, indeed as one of the crofters said, wryly, afterwards, 'The cubs *should* have been thriving, they were being reared on the best of mutton!' To any sheep owner in fact the tufts of scattered, curly wool, the legs, the part-chewed skull of lambs, was enough to cause universal condemnation of the entire vulpine tribe. Mind you, lamb's wool can blow over a wide area and each lamb has *four* legs so that, unless one does a very careful check, lamb remains are exceedingly prone to human exaggeration. When, by dint of burrowing and potholing in the malodorous depths, I got together all the identifiable bits of lamb I had just enough for *four* lambs, though there might well have been other remains underground. There was also the entire framework of a yearling red deer, undoubtedly found dead by the fox and taken in piece-meal to the den. The hind pad of a rabbit and the wing of a grouse was there also, and the nearest rabbits I knew of were three miles distant. The last bit of fox prey I unearthed was the head of what had been a newborn deer calf; pathetically, its wide open, curly eye-lashed eyes still had the blue tinge of extreme youth, cut short prematurely.

My allies, John and Iain, who were only going to wait out until midnight, then had a council of war and we settled on our ambush locations. About to start off for my spot I happened to glance skywards (a habit in eagle country) in time to see first one eagle, then its mate glide high above us, right across the width of the coire without even a wingbeat. Going to roost, I thought, and indeed one of the pair did land on the high sheer cliff of Tom na Gruagach (Hill of the Goblin) while its mate curved back majestically on wideset wings and returned high over us.

By 10 pm we were tucked away out of sight in our positions around the den. I had a near ideal position, ensconced behind a huge rock the top of which, covered in deep soft moss, proved just the correct height as a rest for my .243 rifle. Shielded on both sides, and to my rear by even larger rocks, I had a view over a wide area while remaining unseen. The waiting dragged on, and on, and on, with never a sight or sound of a home-coming fox. Slowly the evening light faded; a distant grouse or two called before settling for the night. Perched in my lofty eyrie well away from sights or sounds of humans, I was utterly divorced from human existence, surrounded as I was by the savage grandeur of this spectacular coire. I was, too, on an ages-old assignment, old as the time from which man first began to keep domestic animals; man against a wily predator which was competing, unwisely, for his flocks. Directly ahead of me, looming up away across the coire, was the high black precipice of Tom na Gruagach; to my right hand was the rock slice which formed the Eag Dhubh of Sgurr Mhor. Very distant there gleamed a segment of Loch Torridon, sombre blue at first then a sombre grey in shade. An awe-inspiring, nay, a downright intimidating place, for in that coire, even in bright sunlight, one always felt quite insignificant, a tiny intruder.

I took my eyes off the emptiness of the heather flat which skirted the cairn to glance up at the equally blank slopes above it. When I swept my eyes back to the now dim heather flat it was no longer empty. A grey-white blur was on the further edge of the flat. Sitting up serenely on its haunches, looking towards the den, was a fox, its grey-white chest towards my position. Wraithlike it seemed and wraithlike it had materialised without any sound to herald its arrival. It looked a terribly tiny target, at 130 yards away, yet the crosshairs of my telescopic sight showed very distinctly on that silvery chest. Quelling my misgivings (I had after all come out here on a punitive task) I steadied my rifle on its mossy rest and squeezed the trigger. The silver grey chest fell over sideways without either twitch or involuntary kick of limbs, to my heartfelt relief.

I did not venture forth immediately since there remained a faint hope that the mate of the fallen fox might yet show up. At midnight the shadowy figures of John and Iain appeared and we groped our way out to the dead fox. It proved to be the vixen, an oldish fox with one fang broken short, though she was well-nourished for a hill fox weighing 14½ lbs.

My allies went home to bed and I settled down, in my sleeping bag for that period from midnight to 3.30 am, when the light would be too dim to see or to shoot. Odours from the den wafted to me as I dozed, a potent mix of acrid fox stench and of decomposing prey. From my recumbent position my skyline was now dominated by the distant ridge of Liathach, etched in silhouette, clear and sharp against a pale, translucent colourless sky. Breathtakingly austere and aloof it looked on that lovely night with its serrated, lengthy ridge soaring to meet the few stars which twinkled wanly in the sky. About 2.30 am I jerked awake out of my uneasy doze to realise that, far away to the east, beyond Ben Eighe, the sky was just assuming a faint wash of rose-pink. Somewhere high above me a ptarmigan croaked a sepulchral, sombre greeting to the new day. Moments later the much cheerier gabble of cock grouse warned me that it was time to be up and on the alert.

Reluctantly, I writhed my way out of my comfortably warm sleeping bag and took up sentry duty again. The huge cairn ahead of me gradually assumed colour as the light grew stronger. A tiny but engagingly perky brown wren quite suddenly burst into deafening song by my ear and minutes later flitted to perch on my rifle barrel. That was all I was to see from my break of day ambush point, for the dog fox, warier than the vixen, or simply still hunting prey, just did not show up. I left at full daylight to make my way home through the very tangible stillness of the surrounding hills broken only by the whisper of the heather against my boots. Another fox family which had broken man-made laws, in dining too well on the lambs of crofters, had met inevitable retribution. I sometimes ponder on an epitaph for the genus fox. Something like this comes to mind: 'They were admirable parents, excellent and adaptable providers of food and they cared for their family well.' Cause for praise or for condemnation?

As you may imagine, in such fascinating new country, our first year at Torridon went by very quickly. There was much to be accomplished besides getting to know the people, the hills and in helping in fox control.

To provide help for visitors, there was a tiny wooden hut which was staffed in the summer. This we had replaced, with a more attractive, bigger, more roomy building, sited at the road junction to the villages of Shieldaig and Diabaig. Interestingly, both of those names are of early Norse derivation and mean, respectively, Herring Bay and Deep Bay. This new building, attractively sited on the edge of a Scots pine clump, was given the rather elaborate title of 'Countryside Centre'. Yet that title was not inappropriate for, inside, it was divided into two rooms, each of which dealt with countryside matters. At one end we had an exhibition room, of Highland wildlife, with both exhibits and photographs of my own, for my second ruling passion in life was the photography of the Highland scene and its wildlife. Later we augmented this with probaby the very first audio-visual display in the Highlands, the theme of which was the countryside of Wester Ross. The 'business' end had the information desk and also charts, maps and reference books dealing with all aspects of the area. Although we were not a Tourist Board office we nevertheless helped out those locals in our area who supplemented their income by providing bed and breakfast and those visitors who desired such accommodation. Our function was to help both locals and visitors and to, in some small way, educate our visitors to the way of life at Torridon, to send them on their way with a fuller understanding than they had arrived with. To this end also, since relatively few people in Britain realise that 'the balance of nature' had, centuries ago, been eliminated by mankind, I formed a deer museum down by our house. This personal venture was fully supported by NTS, and became housed in a beautiful white-tile-walled, one-time dairy. This was in no respect a trophy exhibit; it was designed to show the natural history of red deer, their pattern of life in the Highlands and the very real need for annual humane and informed control, by deerstalking. Our Highland red deer have nowadays *no* natural predators when adults to regulate numbers, such as the extinct wolf, bear and lynx. The owners of any land carrying a population of red deer have therefore to bear this responsibility on their shoulders. This type of control, or regulation of numbers by deerstalking, is designed simply to keep an annual check on the seasonal increase of new-born deer to the herds. This form of deerstalking is *not* a sport, it is a management of the population to their own advantage. There is, in fact, *no sporting stalking* allowed on any NTS mountainous property.

Behind my house we had some spacious pens built in anticipation of any wildlife casualties we might be asked to deal with. As word got around, these were seldom vacant for long. Since Torridon was essentially an out-of-doors walking area we later set up a summer programme of guided walks to show visitors some of the attractions of the outdoors.

We found the indigenous people of Torridon friendly, courteous, helpful and welcoming. I had been born in Fort Augustus when it was a small attractive Highland crofting village, in pre-hydro electric days, when candle and paraffin lamp were the norm. The smell of a guttering candle or a leaky paraffin lamp is still evocative of my childhood. This was in the early 1930s when every crofter kept a milking cow, hens, a few sheep, perhaps a pig for fattening for a winter kill, and, if he could afford it, a horse or pony. The croft, however tiny, was worked for potatoes, turnips, hay and corn. I had an uncle who was a crofter and many of my school fellows were of crofting families. It was a hard, exacting way of life; very seldom was there spare time for any member of a crofting family.

By the end of the Second World War, Fort Augustus was changing.

When improved standards meant stricter regulations, in byres and dairies, in all milk production to guard against tuberculosis, the death knell was nigh for the crofters' milk cow. The introduction of subsidies for beef cattle and for hill sheep meant a further change of emphasis in crofting. Hydro-electricity schemes in the 1950s meant more changes. By the time we went to Torridon, Fort Augustus was no longer a sleepy crofter-orientated village.

Transferring to Torridon in 1969 we found that the leisurely, almost *mañana*, way of life still prevalent there was appealingly reminiscent of my childhood in Fort Augustus. This is not in any way intended to be derogatory – I personally found it very refreshing to see small fields of sweet-smelling meadow hay still being cut by the scythe, and to hear the corncrake, long gone from Fort Augustus. There were disadvantages of course! The weather on the west coast, bulked large in these; it was very often wet and windy. Against this it was seldom very frosty in winter. In 1969 there was little form of public transport to get in or out of Torridon. There does exist now a daily post bus service from Torridon to Kinlochewe with connection to Achnasheen. Achnasheen was, at twenty miles inland, our nearest railhead. The favoured big shopping centre then was a two-hour car journey away, over single track roads, to Inverness. A vehicle of some kind was a necessity at Torridon.

Similarly the nearest accident or casualty hospital was in Inverness while the fire brigade was in Inverness or Dingwall. We had three major fires in the area in our twenty-one years at Torridon. In each case, even though we had a voluntary firefighting team locally, the building concerned was burnt out before the fire brigade arrived. We had one postal lift in the day and one postal delivery, at 3 pm or later.

Social life was a little lacking. We did have periodic dances in the village hall and great dances these were, with Strip the Willow, Eightsome reel, Highland Schottische, St Bernard's waltz etc. going until the early hours. Nowadays, I shudder to write, they have discos! We had a small but enormously enthusiastic badminton club based in this hall. Our hall was low-ceilinged which meant that we had to play a fast, low game. In turn this meant that when we played a neighbouring team, say from Gairloch, which had a high hall, we invariably lost in their high hall, but conclusively won in our own low hall. When we went there this hall was lit by gas mantle but later, NTS paid to have the electricity put in. If, during a rally, you hit a gas mantle, then a let was scored and the game halted until the disintegrated mantle was replaced. We paid a very moderate fee and some of us keener members played on until 1.30 am of a winter's night. Tremendous.

Advantages, priceless advantages, were the spaciousness, the clean air, the peace and tranquility, the lack of vandalism and of *aggressive* drunkenness. Oh yes, whisky was almost universally drunk, too much so on occasions, but even dead drunk there were seldom rows. One knew everyone in the area (I believe a census gave one person to one square mile as the Torridon population) and most were only too willing to be helpful. These advantages are nowadays described as 'the quality of life'; we had quality of life in abundance at Torridon and I am sure that it was that, as much as the superb scenery and the challenging hills, which attracted people back year after year to the area.

The fact that our doctor at Torridon had around 300 patients to cope with was also an advantage; there was no need for appointments. Our nearest policeman was based ten miles away at Kinlochewe and was definitely not overworked. In point of fact, nowadays the nearest policeman to Torridon is now based at Lochcarron, twenty-one miles away.

It has been fashionable to call remote areas of the Highlands, as Torridon, 'disadvantaged'. In twenty-one years at Torridon I met no one, who lived at Torridon, who believed that he or she was 'disadvantaged'. We certainly felt that the balance was well weighted in favour of Torridon.

2. The People of Torridon

As with most sea loch areas of the West Highlands the sea loch of Loch Torridon is the focal point for the small communities throughout the area. In bygone days there were three reasons for this; the communities got a great deal of their sustenance in fish from the actual loch; this same sea loch was the main means of communication between each community when roads were bad or non-existent and thirdly such mediocre 'fertile' land as was available was generally concentrated around sea-level, washed down over the centuries by rivers, burns and weather. The indigenous people of the area were then fishermen/crofters with the emphasis on fishermen. At one time every household had a boat, or access to a boat of some kind, so as to utilise the ever-present fish of the sea loch. In the living memory of the older people of Torridon the sea loch used to be incredibly rich in fish, in utter contrast to today when modern improved techniques in trawling are sweeping the sea beds bare. Even the vast sea beds are finite; sustainable perhaps if wisely used but quite unable to sustain the ceaseless depradation of 'improved' methods in fishing them. In those bygone days it was not always necessary to employ a boat to catch enough fish to supply a household.

Opposite the villages of Annat and of Fasag there are sandy bays which shoal rapidly; on each of these bays there still exists the ruins of what appears to the casual eye to be a low dry-stone wall which curves outwards in a semi-circle and thus encloses an almost circular area of each bay. These walls had at one time been much higher and at high tide were quite covered over. Fish in quantity came over the wall as it was covered at high tide and many were left stranded on the sand in turn as the tide receded. The walls, of dry stone construction and thus full of cracks and crevices, drained freely. It was an extraordinary simple yet enormously effective way of catching fish but it depended on a ready supply of fish. The Gaelic term for this type of communal fish trap was 'cairidh', (pronounced 'carry') and by 1969, while their ruined walls were still in evidence, they were no longer in use. At one time they must have afforded rich harvests for the folk living beside them. Their use is said to have been common to the entire Western coastal area of the Highlands wherever there were suitable bays. I was told of occasional catches of such magnitude that the surplus was spread on the croft fields and dug in as fertiliser. I was also told that the estate head keeper always attended at such a harvest to ensure that all the salmon and sea trout left stranded were 'rescued' and thrown back into the sea, these being regarded as game-fish. Apparently there was also an underkeeper who was more partial to the taste of the sea trout than to the principle that these were all the laird's fish. Whenever opportunity offered he used to stuff a sea trout or two down the waistband of his generously-cut plus fours. Fine, but the journey home afterwards entailed a two-mile cycle ride with this wet, slippery and fishy leg-harness. I wondered if he ever tried it with grouse. It must be realised that most west coast communities, island or mainland, did not look upon it in any way as criminal to take a share in the sea's crop of salmon or sea trout. There was another story, of a well-known worthy of bygone days, caught in the act of netting a river mouth, in a remote bay, at dead of night, by a single keeper, said to him 'Did anyone see you come away down here, Donal'?' The lone keeper replied 'No'. 'Well' said Murdo, laying his hands significantly on a wooden club lying next to him in the dinghy 'it is certain that no one will ever see you return if you do not go away and stop pestering me'.

Annat was the smallest of the villages in our area and lay at the head of the estuary of the loch, under the shadow of Beinn a' Eaglaise (Beinn of the Church). Near to Annat was the former Victorian shooting lodge of Beinn Damh (Ben of the Stag) a magnificent building of red sandstone in a magnificent setting. By 1969 it had been converted in use to that of Loch Torridon Hotel. Annat, sited facing north and below the bulk of a hill, had the sun cut off yearly, from November until the following February. On the other hand, Fasag and Inver Alligin, sited below the much higher mountains of Liathach (the grey mountain) and Beinn Alligin (the jewel mountain) were not at all bothered by their shadows since they were sited facing south. There is surely moral or principle there which no doubt some of our forefathers realised – in mountainous country build your house to face south or southwest.

While talking of houses I was told by Derek MacLean, Inver Alligin, that his father, a stone mason, had built their house, (a handsome building of Torridonian sandstone called 'Broom Cot-tage'), in 1926, for a total cost of £173. He said a major expense was the sum of £3, for ten wooden doors. Beside this house he had a large barn which was beautifully built entirely of dry stone, whereas the house was mortared.

Fasag was the main village in the area owned and administered by the National Trust for Scotland since it held, eventually, the one surviving general purpose shop in that area and the main post office.

By some visitors it was mistakenly called Torridon, but to the locals it was always Fasag. Torridon was the title of the district, or of the estate, not the village. This tiny village was crouched, as between the paws of the couchant lion of Liathach, at its very foot, a thin white line of human endeavour, puny, irresistibly puny, in its impact in that mighty landscape.

West along the loch side, roughly midway between Fasag and Inver Alligin, is Torridon House and its satellite buildings and perhaps a mile west from this is the church which served Torridon and which had been built by a former proprietor of Torridon estate, Duncan Darroch, in 1887, 'in behalf of the majority of the people who may adhere to the Presbyterian form of worship.' Many of the then community of Torridon, listed as 'crofters and cottars on Torridon estate' contributed to the cost of building the church, either in cash, or by a day's work. In a list of these, Alexander MacKenzie, merchant gave £1, while Donald MacBeth gave 1/- (5p) and Kenneth MacKenzie, Rechullin, gave three days labour. These three names are just picked at random from the list referred to, which is substantial.

The former path or track along the shore of Loch Torridon led to Torridon House, thence to the chuch, then continued to Inver Alligin, a charming village larger than Fasag. It is more kindly situated too, being scattered along the lower slopes of Beinn Alligin rather than on sufferance, as it were, on the extreme lower edge of the steeps of Liathach. Wester Alligin, as indeed the name implies, was yet further west. Rather pathetically, Wester Alligin is dotted with the ruins, of Torridonian sandstone, of former tiny croft buildings and with small fields grown over by invasive rushes.

If one carries on along the single track access road which winds high above the Alligin villages, one eventually comes to the road's end, at the sea again. Here is the village of Diabaig, well situated in its deep sheltered bay. This too was a typical crofter/fisherman village coming under the general term of Torridon area but outside the National Trust for Scotland's ownership. Diabaig was nicely sheltered from the east but wide open to seaward where, on the horizon, the Outer Hebrides were visible on a good day. The entire area was largely Free Church orientated in religion. The nearest Roman Catholic church was in Beauly or in Dornie, both around fifty miles distant. Many crofters who provided bed and breakfast facilities would not take any customer who came on a Sunday and, if someone arrived on a Saturday they were expected to stay over until Monday, i.e. not to move out on the Sunday. It was not quite as fanatical as some of the islands. I recall being asked, while on Skye, not to travel or to use the ferry on a Sunday. I once heard an islander tell how his granny chased and caught her cockerel every Saturday night and clapped him under a creel until Monday morning so that he could not sinfully consort with his hens on the Sunday.

By 1969 the school at Fasag served also Annat, the two Alligins and Diabaig. Ample, you may feel, yet at one time both Inver Alligin and Diabaig had their own separate schools. Even the school at Fasag held only nineteen pupils in 1991 when we left Torridon, a number sufficient to ensure against its closure, however.

I referred earlier to the crofting way of life at Torridon. In 1969 the crofters there still made their own hay, despite the ever-present handicap of West Highland weather. They cut it with the scythe and often dried it by hanging it on the fences of the fields. The biggest tractor in the area among the crofters was the tiny, grey-coloured Ferguson which lacked any cab or protection from the weather, a tractor long outmoded in the Fort Augustus area. Peat was still being cut by some families for fuel though this was supplemented by coal from Inverness. Sheep were clipped or sheared in the old way using hand shears although machine-clipping, usually by contractors, was taking over inland, on the bigger sheep grazings.

You may think that an apt way of describing this way of life is 'backward'. I did not; I relished it!

The predominance in the surname of MacKenzie in the Torridon area testified to the days under the clan system when the area was under MacKenzie chieftainship. Less easy to explain was the favouring of three Christian names in the area; Donald came a clear first, followed by Murdo and Finlay was third. Here it was that the old Gaelic way of amplifying a person's Christian name by his vocation, or the name of his house, as Murdo the Post, or Donny 'Bayview', became of value. There were no less than three Donald MacKenzies as neighbours in Annat – they were known as Donny 'Barnfield', Donny 'Bayview', and Donald Hannah. Incidently all of the indigenous folk of the area spoke Gaelic with relish and English of necessity while many of the young folk also spoke Gaelic. The surname numerically second in the area was MacDonald while there were a fair number of MacLeans and MacLennans. In 1969 we added the MacNallys.

More than one of the older folk told me that in the area of Loch Torridon called the Narrows, just beyond Wester Alligin, the coalfish were at one time so large and so packed together on the surface that you could walk across the loch on their backs. The local fishermen could throw out large triple unbaited hooks with the confident expectation of catching a fish and hauling it into the boat. Exaggeration, no doubt, but there was surely a sound basis for the colourful anecdotes.

A yarn more than once repeated to me, of these distant days told of a sea-eagle being found, drowned, its talons still firmly fixed in the back of a huge coalfish. The coalfish had dived, deep, and the eagle had not been able to extricate its firmly imbedded talons and so had drowned while the fish also had, later, succumbed. Skye is just across from Torridon and it was on Skye that the last known sea-eagle breeding pair nested, prior to their becoming extinct as a breeding species in 1916.

There were huge cod to be caught in the loch also. There was a yarn about a crofter's wife who was left with the unsavoury task of getting rid of four newly-born but unwanted collie pups. She made the mistake of using a stout paper bag to put them in, tied and weighted this, and consigned it to the loch. A big cod was caught shortly afterwards, away on the other side of the loch, and inside it were found the four dead puppies.

Large plaice were common inshore, so common that a glass-bottomed, square wooden box, held just below water level from the stern of a dinghy, was used to spot them on the sandy bottom. They were then speared, using a crude, iron, barbed trident, mounted on a fifteen-foot pole. I rescued one such trident head while I was at Torridon and have it yet. It is very similar to the trident which was, at one time, using in spearing (or leistering) salmon, usually by torchlight, in the Highlands. These plaice, incidently, were, on average, the size of a large dinner plate.

From those anecdotes there is no doubt in my mind that the sea loch was immensely important in the lives of the communities who lived on its shores. When actual fish were scarce there were always shellfish to fall back on – one local told me he could never face shellfish as an adult – he had had to eat too much of it as a youngster.

Our nearest neighbours at Torridon were Jean and Bob Robertson who lived in the farmhouse next door to the Mains. Bob was from the north east and a harder working man I have never met. He was then endeavouring to make the farm land at the Mains pay its way by working this single-handed. If any man on Earth could have done this, Bob could have managed it, but in the west with weather and long distances to markets and suppliers against him, he was beaten in the end.

Donnie MacDonald and Morag his wife ran the one and only shop in Fasag. Donnie, inevitably, was better known as Donnie the Shop, or, more obscurely, as Donny London. This latter sobriquet was because his father had at one time been a policeman in London. There were occasions in the summer season when he was addressed as Mr London by visitors from the South. Morag was small, dark haired and impulsive, usually smiling, but with a quick temper when she needed it.

It was from Donny that I got an urgent 'phone message one Saturday afternoon. 'Do you want to see the hen harrier?' he asked (there had been a female hen harrier in the area for a day or two). I, of course, replied 'Yes'. 'Right,' he said, 'come over to my mother's, I've got it in the dustbin there. It attacked my mother's parrot and the parrot gabbed it by its leg. Come on over.' I went hot-foot. A chance to see the hen harrier, confined in a dustbin or no, and to see this redoubtable parrot which had fought back was not to be missed. Donnie's mother lived across the road, opposite the shop, and the parrot was an African grey, renowned for its ability to emit a piercing wolf-whistle at any female visitors to the shop. It had the freedom of the garden and had been there when the attack came.

Alas for my expectations! I accompanied Donnie to the dustbin where he cautiously lifted the lid. Inside there squatted an exceedingly dishevelled common buzzard! 'Sorry, Donnie' I said 'your hen harrier is a buzzard.' An easy mistake, for a female hen harrier is superficially like a buzzard to the casual eye and human nature always plumps for the rarer species in identifying species of birds. Donnie then elaborated on the parrot *versus* buzzard affray. 'My mother,' he said 'hearing the parrot squawking and the buzzard's shrill yelps, rushed out and grabbed the mixed up bundle of animated feathers, then dropped it when she realised what she held. Our two terriers arrived and joined in the fray until they were grabbed and shut in. I arrived then and managed to force open the grip of the parrot's massive beak on the buzzard's leg. Then I put the buzzard in the bin and phoned you.' The indomitable parrot looked quite immaculate and unworried, in contrast to the demoralised buzzard. At Fort Augustus I had recorded voles, shrews, young rabbits, weasels, moles, fledgeling birds, even carrion, such as the afterbirth of a blackface sheep as buzzard prey. I had never thought to record an African grey parrot as potential prey. This parrot was definitely destined to go into Torridon legends for its David and Goliath act. Whether by accident or design, when the buzzard had swooped and grabbed for it, with out-stretched talons, the militant parrot had neatly dodged and as neatly grabbed one outstretched, taloned leg, in its massive, nut-cracker beak, and held on grimly. The evidence of the power of its grip was there in the bloodstained wound across the yellow of the buzzard's leg.

I told Donnie that I would take the buzzard home, do what I could for its injured leg and injured feelings, and turn it loose in a day or two, sufficiently far away to ensure against a repeat assault. Privately, I was convinced that nothing on this earth would ever persuade that buzzard to again attack an African grey parrot.

Since Torridon is a Highland deer forest area many of the locals had, at one time or another, been employed on these deer forests, as ghillies or helpers, either on Beinn Damh forest or on Torridon forest. The network of hill-paths on these forests had originally been constructed by estate staff in the 1800s so as to enable ponies to go to the hill in the stalking seasons, to carry guests out and deer back from the hill. Exceptionally well-constructed, these paths were in those days maintained annually, by the estate ghillies and stalkers, in early summer. Drains were cleared, soft bits in the paths were bottomed with stone, landslips were cleared. This work, on these paths, went on until the Second World War at least and has left a legacy of paths which most hill walkers are grateful for. Their maintenance on NTS hill properties is nowadays an annual job of work.

Finlay, an elderly crofter in Fasag who kept a small flock of Cheviot sheep, told me that in his day the estates provided much of the work available. Torridon estate for instance employed four ghillies and two stalkers. He remembered a Lord Knutsford who had had the stalking tenancy of Torridon in the early 1900s. He had been walking in Torridon house woods when he quite unexpectedly came on an elderly crofter wife from Rechullin who, head bent, was gathering dry twigs for kindlings. 'Good morning' he said, in rather autocratic manner. 'I am Lord Knutsford, the tenant at Torridon house.' The old lady straightened up and replied 'Ach well, my lord, there is an even higher Lord watching us from up yonder,' pointing to the sky, and bent down to her twig-gathering once more. Finlay's house was directly below Liathach, so near to the screes and rocks of the steep face that one February he actually watched from his back door a pair of foxes at their mating on the hill. I must admit he did not appreciate the sight and, having no gun in the house he phoned me, to find that, luckily, I was not in. From how many houses in all Britain could you hope to see vixen and dog fox on their honeymoon?

Finlay told me of a stalker who, in the days prior to telescopic sights, had a gentleman out who always required the stalker to whiten his rifle's foresight with a little piece of tailor's chalk which he carried for that purpose. The chap concerned was not a good shot and the whitening of the foresight was to enable him to see this more easily. In taking his 'gent' into a stag the stalker forgot to apply the tailor's chalk and stag was missed. Predicably, the discomfited gent rounded on the stalker and gave him a really nasty tongue-lashing, blaming the lack of the chalk for his miss.

Later that day they were fortunate enough to get a second stalk. The stalker was ostentatious in applying the tailor's chalk, liberally. The second stag was also missed. 'Well,' said the unfortunately outspoken stalker 'it wasn't the lack of chalk that made you miss that stag!' These were hard days, the years between the Wars, and that remark cost him his job.

He told me also of how he had been out with another chap whom neither himself, nor the stalker, knew was an estate factor. In one of the high coires they came on a big clump of dockens. 'My goodness', remarked the factor 'surely it is most unusual to find dockens away out here, high on the hill.' 'Ach,' said the stalker, 'factors and dockens, they'll thrive anywhere!' Factors, generally, have, I'm afraid, never been popular in the Highlands.

Another Finlay, but from Annat, had been a stalker on Beinn Damh forest. He had had a daughter of the owner out for her very first stag, a red letter day for most youngsters. To his disgust they saw nothing in the way of a good stag all day. They were returning home when he saw a young stag with a few hinds. His young protégé got it. Finlay was rather ashamed when he went up to gralloch it, it was even more insignificant a stag than he had thought. The youngster was, delighted, knowing relatively little about red deer. She insisted on coming out next day to help take 'her' stag in on the pony since they had had to leave it out that night. The ponyman employed in that season was Sammy, young, fit and strong, and he and Finlay, with the pony, arrived at the dead staggie before the youngster and another ghillie. Sammy, keen and eager to show his fitness for the job, almost sniggered when he saw the size of the 'stag', and grabbing it, was swinging it up easily, on the deersaddle of the pony. 'Drop it, you bloody young fool' hissed Finlay, looking back to where the others were just appearing, 'drop it. Make it look heavy. Wait until Donald comes up to give you a hand and then pretend it takes the two of you with all your strength to load it on the pony.' Donald arrived a moment or two later with the youngster and the 'stag' was then loaded with due heaving, straining and exhausted grunts. Honour satisfied, the cortège then made its way home.

Finlay was very much a traditionalist and to his mind the best treatment for his hill boots was to treat them liberally with stag's fat, rendered down appropriately. It was quite a protracted, laborious job, and, satisfied with his efforts, one night, he laid the boots aside and went to bed. He had then a young collie who stayed in the house all night. Absolute disaster ensued. The young collie had strong teeth and an all-consuming hunger. All that was left of Finlay's stag-fat-anointed boots in the morning were the iron-shod soles and heels.

Sammy the pony man referred to above lived across in Newton, beside the River Torridon. His father was an exception to Torridon surnames, a Thorburn, who had met Sammy's mother when she had been in Glasgow. Jim Thorburn had lost an arm in the First World War yet it was hard to realise that he had only one arm so well did he manage, even to driving his car. Sammy's mother, Johann, was a native of Torridon; her father had been a stalker and they had had a croft. When down in Glen Torridon, one late summer evening, to fetch the croft cows which had strayed, a couple of local lads were spotted dragging a dead stag. They hid this under bracken near the road to fetch after it got dark and then made off. Taking the cows home, the father was informed whereupon he made off down the glen and took the stag home. The lads had been recognised but no further punitive action was taken; the loss of the stag was believed sufficient punishment in these circumstances.

Sammy's wife, Anne, was from Lewis and when we went to Torridon, in 1969, they lived in a residential caravan, on the croft ground, with their three children. By the time we left, in 1991, they had built their own house, also on the croft ground. Sammy himself was a natural genius with anything mechanical or electrical; if anything of that nature went wrong, then Sammy was the man to turn to.

It was Sammy's mother, Johann Thorburn, who told me of the old belief, current in the area, of the cure for epilepsy which entailed the use of the skull of a suicide as a 'cup' from which to drink from a running stream. Such a skull was kept for this purpose under a flat slab of rock, in a hollow on the lower slopes of Beinn 'a Eaglaise, about mid-way between Annat and Newton. She showed this to me and I took a photograph of the pathetic scrap of skull which was concealed there. The 'cure' was this: give the sufferer three drinks of burn water from the skull of a suicide, one in the name of each of the Trinity. The cure is said to be dependent on the faith of the patient. He or she is told that belief in the cure must be total or there will be no cure. The thinner portions of the skull at Annat had long crumbled away, nor was it any longer white, but very discoloured. It is said to be of a Mrs Mary Grant who had originally come from the Lochbroom area with her parents. Towards the end of the eighteenth century she had married a Donald Grant. The poor women's mind became unhinged, and she had to be watched all the time. One day, however, when the pair were working at cutting peats above Loch Torridon she somehow contrived to slip away and threw herself over a cliff and into the sea loch where she drowned. Friends later tried to bury her in the cemetery at Annat but the hard-hearted pious of those days turned them away. Mary Grant's body was then buried in unconsecrated ground which had also to be out of sight of the sea, otherwise, it was believed, the fish would desert the loch. The belief in the efficacy of the cure was exceedingly strong and epileptics came from far and near to be cured. One of the last to use it was a lad from the Gairloch area, in 1944, and, it is said, he became cured as a result of his pilgrimage to Torridon.

The local 'newsman' of the area was called, inevitably, Donald a very hospitable character, always ready to ask you in for a dram if he saw you passing by. If there was any 'news' in the area Donald was sure to have it and to embellish it a trifle, simply to make it a better story. He also had a bit of a reputation, (self-confessed, I admit) of having had a beast (deer) from the hill whenever he had wanted one, when he was younger. Crofters in rural areas in days gone by knew how to utilise every possible bit of a sheep or cattle killed for household use. Black puddings were made from the blood of a fresh killed beast. Donald had heard somewhere that fresh blood drained from a red deer made top quality black puddings. Donald's house abutted on Liathach and one day in winter he spotted a group of hinds on the hill face above his house. He grabbed his rifle and passing through his kitchen he saw a brand new red plastic bucket his wife has recently bought. Black puddings yet on his mind he grabbed at it and away to the hill, stalked 'his' deer, trailing the red bucket, got one, and caught her blood in the new red bucket. I had thought that I knew most aspects of deerstalking but that was a completely new one to me. Donald maintained that the black puddings made as a result were very good indeed but that his wife, thereafter, refused to use her new red bucket.

Poaching of salmon or deer was not recognised as the most cardinal sin so long as it was done 'for the pot'. This stems back to the days before the mid 1800s when sporting deerstalking became fashionable to those with plenty of money and leisure. Before this, I am quite certain, any Highlander so minded had a 'deer from the hill' when he was so minded and equipped to do so. I suspect that, times being hard and the population larger than the bare areas of the Highlands could always provide for, that even a discovered red deer calf, new-born on the hill in June, could end up in the stew pot of a meat-hungry family. In this way an almost constant check was imposed on the numbers of wild red deer which peopled the hills then.

I was told by another friend, Willie, that at one time there had even been a butcher at Diabaig who used to buy stock, as sheep or cattle, locally, when this was available, and slaughter these himself before dressing the carcases out for sale. In winter he bought carcases from a butcher in Dingwall since stock could not be procured locally. At the end of each winter he used to cancel his buying from Dingwall. There were no such things as telephones generally available in Diabaig then so he used to send a wire (telegram) via the Post Office. The last telegram he sent became a much quoted classic in the Diabaig of that day. I quote 'Don't send any more meat; killing myself tomorrow'.

People in crofting areas of the Highlands were much more self-sufficient even as little as forty years ago. A pig was often kept by a crofter simply to kill and dress out for winter keep.

Angus Braebeg was an elderly crofter whose family had lived in the area for generations and he had found employment, as so many others had, as a ghillie on Torridon estate. He too had a great fund of yarns. He also had the copper pot of an ancient whisky still which, he claimed, had been his grandfather's. Angus also told me of a fascinating encounter he had once witnessed while resting, sitting quietly, high on Beinn Alligin, many years before. A movement some way below him caught his eye and he watched a hen ptarmigan lead her brood of seven chicks from an area of grey scree. Almost simultaneously a small falcon, a merlin, he thought, hurtled into view and clutched a chick. The mother ptarmigan at once flew like a rocket straight at the merlin just as it tried to take off with her chick. Cannoning into the merlin she knocked it flying in a dishevelled heap of feathers, causing it to drop her chick. Completely and utterly demoralised the merlin shakily flew away without prey while the ptarmigan and her reunited brood vanished again into the scree.

Murdo the Post was another crofter who, besides being our postman for many years, ran his croft at Wester Alligin. He loved getting away to the hill and since he had a small stock of sheep he had every excuse for this. One day he was spying from the upper slope of Beinn Alligin, looking westwards, with his telescope. A glimpse of vivid rust-red had him steadying his glass and 'yes' it *was* a fox, a big dog fox in beautiful condition, having a leisurely meal from the carcase of a red deer. Even as he watched the far distant fox an eagle appeared in the field of view of his glass and, landing, caused the startled fox to jump sideways in alarm, his meal interrupted. The fox, at the head end of the carcase, raised his two front legs onto it. The eagle stood towards the rear end of the carcase, every one of its pale gold head and neck feathers abristle, it's beak half-open. The fox faced it, head on across the length of the dead deer, every red-gold hair fluffed out so that he looked larger than he was, jaws open, agape with sharp white fangs. The fox was the first to try an aggressive move, advancing a step or two; the eagle, no whit abashed, so far from withdrawing, also advanced along the carcase, wings now half-held out from its body, beak menacingly half-open. Not a sound in all this small drama, for neither fox nor eagle are loud-mouthed in their activities.

The position remained an utter stalemate for some moments while Murdo watched spellbound, hardly daring to breathe though he was so far away. First to break, in the silent duel of menace, was the fox; he took a tentative step backwards and the tension-laden spell was over. He had probably eaten his fill before the eagle arrived; had he been sharp-set he may well have contested the issue longer. Murdo, grinning all over his face, told me 'He obviously didn't like leaving his meal; he went off very slowly, "grinning", white-toothed, over his shoulder at the eagle as he did so. He would be back, perhaps after dark, for another feed; right now it wasn't worth tackling the eagle. I envy both Angus and Murdo those glimpses; hillmen both, and reliable; what interested them on the hill interested me, which was why we blethered to each other.

Sandy, Coulin, was yet another hillman I enjoyed a blether with. A heavy-set man he was yet an excellent athlete and many a memorable tussle we had on the badminton floor. Sandy had never heard of the action of ages-old glaciers on Glen Torridon nor of glacial moraine. He was once walking through that area of hill which lies above Loch an Iasgair (the loch of the fisher), and almost opposite to Coire Dhubh (the black coire) called, in Gaelic, Coire Ceud Cnoc, and in English, the coire of a hundred hillocks. This is a very apt description for an area full of conical heather-clad knolls all cheek by jowl with one another. In many parts of the Highlands these are attributed to fairies and are sometimes called 'fairy knolls'. Sandy was third in a party which included a Nature Conservancy chap in the lead, a visiting chap to the area next, then Sandy. The NCC chap was overheard by Sandy informing the visitor these hillocks were the result of glacial moraine, centuries ago. Sandy, blunt as ever, was then heard to mutter, as if to himself, 'Bosh! everyone knows they are just big mounds of gravel, grown over with heather.' A rose by any other name.

The utter informality and lack of petty restrictions in village life at Torridon appealed immensely. As an instance, our eldest son, also Lea, was twenty-one some years after we went to Torridon. We thought that we would have a party at Torridon for him and also as a sort of thank you to all these people in Torridon who had made us so welcome. We were able to hire the village hall for a very modest sum and we organised the appropriate refreshments ourselves. As far as I can recall the party began around 10 pm while it was yet daylight in June; it ended about 6 am the following morning, daylight again.

There were one or two minor hiccups. A chap from Diabaig, reputedly a simple soul, turned up, walked straight up to Margaret, my wife, and said 'I wasn't asked but I didn't take offence, I came all the same.' Nobody was nastily drunk though quite a few were happily hazy. What liquor we drank we danced off. The whisky ran dry at last and that did not lessen the enjoyment one bit.

There were a few other folk living in Torridon who were also incomers. Most of them had integrated well and, in doing so, had, in due course, contributed to the community. Charlie and Priscilla Rose had forsaken London to come and live in Inveralligin. Charlie, a skilled rock climber and mountaineer, now organised and led our mountain rescue team while Priscilla organised evening get-togethers of craft work for the ladies each winter. Peter and Jean Grensted also forsook England for Torridon where Peter became the local expert in television in our area and indeed pioneered good reception for us. Peter and Jean also took over the responsibility of looking after the village hall, much to everyone's satisfaction. Dan Livingstone had come to live at Torridon from Rutherglen and he also took full part in community activities including the mountain rescue work. Neil and Irene Rieley came from the Glasgow area and looked after the magnificent new Youth Hostel which was built in 1975. This is a Grade A hostel, of the Scottish Youth Hostels Association, and it has served the area well, being full to overflowing each summer. Jimmie Munro had come from Easter Ross to Wester Ross. He and his wife, Cathy, ran the post office in Fasag when we first came to Torridon. A great character, Jimmie had been a policeman in Kintail for many years and was a dab hand on the fiddle. He also grew the finest roses in the whole of Torridon – each summer his garden was a symphony of colour and scent. Jimmie had a mordant kind of humour. I passed him one day while he was leaning on the roadside dyke, gazing down at the numerous rabbits at the bottom of a field at the loch's edge. He turned to me, absolutely poker-faced and said 'There are one thousand and one rabbits down there' and then, immediately, as if to counter any scorn I might pour on this assertion, he said 'I've counted them!'

May Carswell had come from Edinbugh and she ran a bed and breakfast house in Alligin where her baking and cooking became a local legend. Her way of driving on our single track roads, with the only automatic transmission car in the area, also became something of a legend. Dr Bill Turner was our doctor for years. He had travelled the world in his profession, including a whaling station in South Georgia, before he and his family came to Torridon. A quiet, unassuming man, he was an excellent doctor who became a friend as well as our doctor. When he died, tragically, in harness as it were, of a heart attack, simply because he never spared himself, he was sorely missed. We employed, wherever possible, people living locally, in our Countryside Centre each summer season. In turn we had Jane Radford, Alex Mary Connell, Katherine MacKenzie and Ruth Thorburn, all, in some degree or other, enhanced and improved the Centre with ideas and suggestions. In fact the only minor disasters we ever had with seasonal staff was when, in emergency, we had to employ someone from outside the area altogether. Nor am I being starry-eyed in this assertion; it is simple fact.

I cannot possibly end this chapter on the people we met in Torridon without mention of Alice Maconochie who, at that time, ruled Inverewe gardens with benevolent autocracy. A tiny, precise lady of big heart and enormous energy she initiated Margaret and me into the mysteries of NTS administration. We will always remember her with affection. She was aided then by 'big Jean' Gibb who, in contrast to Alice, was of robust stature as her affectionate sobriquet implied. She remains a friend; we were at the Highland reception when she married Sandy Lindsay and I remember yet, at Dundonnell, in Wester Ross, the masses of strawberries and cream I ate, in May.

3. The Mountains

It has been said of Torridon that Nature lavished beauty but ignored productivity; that could also be said about most of the West Highlands. I imagine that it depends what you are looking for; one would not go to an intensively industrialised area of Britain to look for scenic beauty, similarly one would be insane to go to Torridon to look for intensive industrialisation.

Of all the magnificent mountains in the Torridon area, Liathach unquestionably reigns supreme; it is immense, it has atmosphere, it has height and it rises, dauntingly steep, from sea level to 3456 ft. Viewed from any angle it has an intriguing and challenging skyline. Its southern face, directly above Glen Torridon, is uncompromisingly hostile, rampart after rampart rising in forbidding rock tiers nearly to its splintered skyline. One felt, on this face, that Liathach was alive, ever alert to pounce upon any mistake. I once watched a stag cartwheel down that face and wished, afterwards, that I had not done so. Linked to my close participation in mountain rescue work it showed me, in stark detail, just what damage could occur to any human falling down that face. Imagination, in reconstructing a mountain fatality, tends to visualise a free fall, accelerating at 32.2 feet per second through the air, the crunch coming at the bottom. Seldom does it occur like that! In the case of the unfortunate stag I saw it was a fast-accelerating series of bone-crunching crashes and whirling rebounds out into space whilst I cringed into myself and yet could not shut my eyes until the falling stag finally came to rest way down the face. I checked afterwards and I do not believe that there was a bone unbroken in its entire body; the antlers which it had possessed I never did find. By contrast the northern face of Liathach was almost fertile and welcoming, with a series of high, verdant-looking green coires. This northern side of Liathach formed the best red deer habitat on Torridon and when we began our programme of guided walks it was into these high green coires that I took visitors who were keen to see red deer. These sightings of red deer, I must emphasise, did not entail a close approach. It was much more rewarding to lie at a distance and watch the deer, truly red in summer coat, standing out against the bright green of summer grass, as they went about their daily life, undisturbed, and to go away leaving them still undisturbed.

The high crags, stretching to the summit ridge, were damp and black, interspersed by vivid green ledges and runners, luxuriant in fern, moss and flower. Most apparent were rose root, mountain sorrel and globe flower while parsley fern, starry and mossy saxifrage were less apparent but, to the discerning eye, ever present. Right up on the summit ridge I found a white variation of mountain thyme and in the unappealing sterility of the quartzite which capped Mullach an Rathain there was, annually, Arctic mouse ear. Nothing very rare, unless you count *loseularia procumbens* (the mountain azalea) or *betula nana* (dwarf birch) as rare, but all were interesting and eye-catching. Liathach had, too, a spectacular coire, Coire na Caime (the crooked coire) which, to me, rivalled the better known Coire Mhic Fhearaichar (the coire of the Farquhars) of Beinn Eighe (the file mountain). Backed by the spectacular skyline of the Pinnacles ridge, Liathach had a series of small lochans and rock outcrops which descended from Lochan a' Glas Thuill (the lochan of the grey hole), and fed each other as they dropped downhill. In turn, Lochan a' Glas Thuill was fed, via a waterfall, white and picturesque when in spate, from a burn coming from the even higher green coire which was cupped between the main ridge of Mullach an Rathain and the Northern Pinnacles ridge. These Northern Pinnacles were so shattered and splintered with rotten rock that they required skilled climbers employing rope-work techniques. So awe-inspiring did this ridge appear that it was seldom ever attempted. Yet I have seen a red deer stag up on this ridge at the rutting time, isolated, splendid in silhouette against the sky, roaring, braggart of voice, while, in the chill of that October day, a cloud of silvery vapour floated away from its uplifted muzzle. Such a sight my camera could not adequately recapture.

Liathach, inevitably presented a challenge. It was always first choice for aspiring youngsters who wanted a guided walk. I set myself to learn thoroughly the various ways to get on and off the mountain, apart from the two recognised routes up the southern face. These two were the very steep track which went up about half a mile east of Glen Cottage and from thence, east along to Spidean a' Choire Leith (the peak of the grey coire) or, as an alternative, the track which left from the westerly wooded clump, by the River Torridon, and wound its way up into Coire Allt Tuill a' Bhan (coire of the white burn) *en route* to Mullach an Rathain (the top of the Pulley). Both of these were accepted as safe, well-marked routes, but steep and arduous, particularly in their final sections to the summit ridge. Behind the Youth Hostel, the Stone Shoot reared its way up to the summit ridge, red-brown in its long, long train of large, unstable blocks of sandstone. This was rather hazardous in the looseness of its rocks; it was accepted as of some value as an emergency escape route *off* the ridge but as a route up to the top it was no longer favoured. Just west of the Stone Shoot, almost directly above the village hall, I pioneered a fresh route which led up the face, steeply at first, then slanted diagonally up through screes, below a line of outcrops, to then swing back east to slant up to the summit ridge near Mullach an Rathain.

The face above Loch Torridon could also be climbed in a couple of places, the first up the east side of the Ard Ghoil burn, while from the west side of the same burn, one could slant steeply up, on a diagonal line, to reach the west end of the ridge at Sgurr a' Chadail (the peak of rest). This latter was well named, by the time you had got up to it, you were more than ready to rest. Mind you, on any fine day, you had ample excuse to pause to admire the views which opened out below you, peering down as into the chimneys of Fasag and across to the hills of Beinn Damph, Shieldaig and Applecross, beyond Loch Torridon. The guided walks were designed for people of differing walking abilities and fitness levels, to take them to areas into which there were no paths and into which they were unlikely to penetrate without guidance. A certain level of fitness was required; hill walking in Torridon is not enjoyable for those to whom walking is a bad word. I began most of my walks with a steep uphill slog so that those who were grossly unfit could pull out voluntarily while there was yet time to direct them on to a safe path, visible below. This system weeded out people painlessly, or perhaps that is not quite the most apt term. At any rate it avoided pointless recrimination if walkers discovered for themselves that they were not fit.

Obviously, too, with a mixed party, one had also to adjust to the pace of the slowest walker. This could be frustrating for the faster walkers but it was a matter of safety on the Torridon hills equally as much as a natural regard for the enjoyment of all the party. The cardinal rule on these walks in Torridon was 'play it safe'; one was responsible for people of varied abilities, one had to come down to the lowest denominator. Ideally there should be a guide at front and rear; since this was seldom feasible one had to impress on all participants the rule that no one at the rear of a line of walkers should allow themselves to lose touch with the party. This rule was absolutely essential in misty conditions.

It gave me enormous pleasure to watch the obvious enjoyment that people had in being taken into areas of the hill quite outside their day-to-day experience. To take, for instance, people from an urban or industrial area of Britain into Coire Toll a' Mhadaidhe, with its primeval aura, was akin to taking them to the moon. There were youngsters who had never even seen a domestic hen in their lives far less red deer in a Highland setting. An army cadet corps of teenage lads from the English Midlands camped with us one November. The chap in charge came to see me for a chat. I asked how the lads were enjoying it. 'You won't believe this,' he said 'the thing which has all of them worried most is the darkness.' Until that moment I had not realised that those people in Britain living in an urban environment never experience complete and utter darkness. To give a youngster their very first sight of red deer, or ptarmigan, or golden eagle and to demonstrate that the apparently empty hills had a life of their own was very rewarding. To infect them with enthusiasm like my own, without leaving them too starry-eyed about it all, was, I thought, worthwhile. There were unpredictable bonuses; a meadow pipit's incredibly neatly-woven nest with its clutch of eggs; a mountain hare, crouched motionless in a clutter of grey rocks; a new-born dappled red deer calf, annual wonder of the June hills; an eagle, soaring, high above us; a black throated diver displaying on an inky-black Highland loch. Unforgettable moments for me, never mind for the townsman.

If I have a criticism of the dedicated hillwalker, or the confirmed Munro bagger it is that so often they walk the hill with only one thing in mind, to reach the top. Living among hills most of my life, this has never infected me. An elderly crofter in Morvern was once asked what the view was like from a high, dominant peak in the area. His reply was typical; 'Well, do you know,' he said, 'I just never ever had occasion to go up there, so I just could not tell you.' An embarrassment of riches?

I was perhaps exceedingly fortunate that in all the years I led guided walks, in weather of quite unpredictable variety, I never had a participant have anything but a very minor accident, such as a grazed knee, for instance. Since those whom I had outranged in age from six to sixty this was quite gratifying. The most alarming accident which ever occurred happened on Liathach. A chap who was in his early thirties, lean and fit-looking, suddenly collapsed, without warning, whilst we were negotiating some scree. He rolled, slack-limbed, down the uneven scree and crashed to a halt at a large rock. Racing down I found he was uninjured but quite unconscious. Some anxious moments later he came round and confessed that he was an epileptic and, worse, he had forgotten to take his pills that morning. He appeared to recover well but that day was ruined for all of us until we got him safely off the hill. Had we been threading along the top of the sheer outcrop his lapse in omitting his daily dose of medicine could have cost him dear.

Mist was the hazard I disliked most of all. Not one of the Torridon hills, with their all-too-frequent sheer rock outcrops, was guaranteed trouble-free in thick mist. I remember one day when I took a party of seven to the top of Mullach an Rathain on a lovely clear morning. Halfway up, the mist came down but we got to the top and, as we had lunch there, the mist cleared. As we finished our lunch the mist came down again, this time thick, woolly and impenetrable. We waited, and waited, but it showed no signs of clearing. The breeze had dropped, my party was complaining of the cold, for clammy, damp mist, at 3000 feet, is seldom warm. I decided to try to grope my way down and led off slowly, impressing on each member of my party not to lose sight of each other at all costs. A little way from the top I encountered five other walkers, who, any port in a storm, wanted to join my party. Not exactly cheered at this added responsibility, I led on to find the top of a steep, narrow, rocky gully which I knew eventually linked up with the Stone Shoot. Down its rocky, confined steepness I went, twelve folk of very mixed walking ability following me. So steep was it, so uncanny was the feeling of complete isolation engendered by the thick mist, that I had to go up and down that line, reassuring people continually, placing individual feet in felt-for toe holds, emanating all the time a cheerful confidence which should have qualified me for an Oscar. It was an infinitely slow, infinitely anxious, infinitely careful descent. A slip or fall by anyone at the top of that line would have swept everyone below off like so many flies. At about the 1000 feet contour the mist ceased, the grey line above us remaining as if marked out by a ruler. Below us, blessedly, was the village hall. Very thankfully, we reached the road at last; my train of walkers thanked me and assured me that they had all had an unforgettable day. I believed them.

I had, of course, to check out the Pinnacles section of Liathach, the ridge walk, which obviously would attract some hillwalkers, to see whether this was suitable for a guided party. By now I had a guided walks programme worked out; the weekly programme was a half-day walk for beginners, really to introduce people to the roughness of hill walking at Torridon. This was on a Monday; on Wednesday I had an all day walk usually into one of the attractive hill coires; on Friday I had a high level walk, to one of the high tops. All of these, of course, depended very much on the weather, particularly the high level walks on which, too, I imposed a participant level of six in the party.

The chance to prospect the Pinnacles ridge came when two lean and fit young students asked if I would guide them over this section of Liathach. The day was sultry, rather too warm in fact, but it was practically windless, ideal for the exposed Pinnacles. We started at the east end, first going up the Coire Dhubh path to a convenient height at which to swing west onto the flank of Liathach. There we angled up, keeping to the top edge of a series of screes, until we reached the gully which leads to the top. Near the summit ridge I found a lovely vigorous plant of a white variety of mountain thyme, the first I'd ever seen of that unusual colour in this flower. We then backtracked along the uncomfortably angled blocks of quartzite which formed the ridge, to the eastern top, Stuc a' Coire Dhubh Bhig (Peak of the little black coire). The two lads lived up to their lean and hungry look and we made good time. Up on top it was sufficiently high for the sultriness of the glen to be cooled to ideal walking conditions. Unfortunately there was that inevitable accompaniment to sultry summer weather, an exasperating blue haze, with the distant views all masked.

From Stuc a' Coire Dhubh Bhig we travelled the shattered quartzite top, that tremendous cone of grey-white quartzite blocks which forms Spidean an Coire Leith. Sixty or so red deer we saw on the way there, well below us, on the high greens of the north-facing coires. The walking on that section of the quartzite blocks ridge was distinctly uncomfortable and one had little chance to look anywhere but at one's footholds. A quartzite ridge is not a walking surface I would choose as a comfortable surface. Indeed this applies as a resting place also for, resting for a spell on the top, one had to search around for a comfortable reclining spot.

Up there we heard a familiar musical *clonk-clonk-clonk* and, looking up, saw that a family of five sable ravens were practically stalled in the air, hovering, looking us over. I hoped it wasn't an augury for our negotiating of the Pinnacles section, for the Highland deerstalker's belief was that the raven can 'smell' the likelihood of a supply of carrion. I did not tell the two lads that they might be in that category, in the eyes of the ravens. Certainly from our viewpoint on top the Pinnacles appeared awe-inspiring. The only other climbers we saw all that day obviously thought so too for they all avoided its challenge. Perhaps they also had a superstition about ravens.

In practice, when I embarked upon the Pinnacles, they did not prove quite as formidable as they had looked. They were difficult, the air of absolute exposure was ever-present and you had to concentrate on every foothold and handhold. There were places where one had to reverse and climb down backwards, the gulf yawning to either side. Above all, it was no place at all for anyone with the least suspicion of vertigo. Equally there was a tremendous exhilaration in being perched on a narrow rim of the earth with everything below you and, before you, on a thread of shattered Torridonian sandstone. I formed the impression that, while it had tricky bits, given reasonable weather conditions and fit, active walkers, with no undue fear of heights or exposure, the Pinnacles section was entirely feasible. Nevertheless you had absolutely *no* margin for error and in adverse weather conditions of snow, ice, high wind or heavy rain, the Pinnacles were to be avoided. I decided that if I ever took walkers across this part of the ridge that I had to be abundantly sure of their fitness first. As always, I had to play safe with the lives of others.

The Pinnacles over, the ravens disappointed, the rest of the ridge, impressive enough ordinarily, became just a slog, up to Mullach an Rathain, where we ate a late lunch. Eating, relaxed, achievement flowing nicely through us, we watched around one hundred red deer, hinds and calves, equally relaxed, way below us, lying on the wide sweep of green hillside, chewing a reflective cud. Nor was this the last bonus for, making a leisurely way afterwards to Sgurr a' Chadail, we discovered a bountiful supply of crowberries, slightly acid and gritty, but refreshing, and, near to them, some juicier blaeberries. Discovering a mutual like for berries, we did both species full justice. Replete, we slanted down to reach the loch at the Ard Ghoile burn. It had taken us from 9.15 am until 5.15 pm to do all of Liathach ridge, and to enjoy it, without rushing it.

A quite unexpected bonus occurred on another guided walk of Liathach one summer. On this particular walk I was able to tip the scales in favour of the survival of a deer calf which otherwise would have perished. The Monadhliaths in Inverness-shire, my former stamping ground, is a range full of deep peat runners and bogs. Many of these peat runners, formed by water action, have become grown over and so are underground, but with holes here and there, dropping vertically, through areas of thin soil, into them. They can be quite a hazard to very young deer calves. These calves, either playing nearby, or just curiously investigating the black holes, can drop down, to become immured below. From just such an underground 'prison' I had rescued a Monadhliaths' deer calf some years before, attracted by the fuss the mother hind was creating, standing over the small hole, pawing at it with her front hooves and calling persistently. I investigated, spotted the calf about four feet down, found a peat tunnel which ran into the underground 'dungeon' and rescued the calf. I was to repeat such a 'mountain rescue' on Liathach.

We had gone to the hill early that day; I had a small group of walkers who were keen to see red deer and, if possible, some recently-born dappled deer calves. The morning was grey and rather uninspiring but it was dry. I had my cameras in my bag and also a small spring balance with which I had been weighing a golden eaglet, weekly. A two-hour walk took us into hind territory; the first two hinds we saw had each a strong, recently-born, dappled calf following them and my companions drank in their youthful charm as they frolicked around the staid mothers whose sole preoccupation was grazing. We saw many more deer that morning and quite a few new calves, but none were in situations where we could get close-up views. We paused for lunch, the clouds were thinning now and the sun came through at last. I led the way then to a commanding ridge and, from its vantage point, spied at the coire below. I'd hardly begun to spy when I spotted a hind with a small dappled calf prodding at her udder vigorously while it fed, greedily. I watched it disappear, after its feed, into a peaty hollow just over a rolling ridge, a hollow which I knew held peat runners and also a well-used peat wallow.

We all crossed to that ridge and scanned the hollow below. Empty it was, or rather it appeared empty for I knew well enough how adept at concealment a young deer calf could be. I began to walk slowly through the hollow, pausing to scan every bit of possible cover. In the lowest bit of the hollow, just where a well-used deer track crossed this was a small, roughly oval hole in the green which seemed to go down, dark-shadowed, into black peat. 'A nasty trap for a deer calf,' I thought, remembering my Monadhliaths experience, then dismissing it. Until suddenly I realised that I had seen a flicker of movement from its depths. Back I went, doffed my rucsack and, going over, peered into the hole. There, about three feet down, black smeared with peat so as to be unrecognisable, was a deer calf, standing, hunched up and cramped, in six inches of wet peat. So narrow was the black-edged hole that I had great difficulty in getting both arms inside it and around the struggling calf. Getting it lifted out was like extracting a cork from a bottle, a 'cork' too which was wet and slippery with liquid peat. Much of this peat the calf contrived to transfer to me while I struggled to free it. The hole measured only about two feet in length and eighteen inches across. Finally success-ful, I scraped as much as possible of the clinging, wet peat from the shivering body of the calf. It was a male calf and, since I had the spring balance with me, I weighed it. He scaled twenty-one pounds. His vigorous, high-pitched squealing brought forth, from an appa-rently empty hill, the anxious lowing call of the mother hind who came racing, hot foot, to the rescue of her calf. Stiff-braced hooves brought her to a slithering halt when she saw us grouped around her calf. Immediately I released the calf; the hind called, low-pitched to her calf again, the calf gave a muted squeal and raced up to her. We watched her lick briefly at his head and ears, still peat black, before she turned and led off her restored-to-life-calf.

Beinn Alligin is always and inevitably seen in the shadow of Liathach and, also inevitably, is compared to it. Most people tend to regard it as a 'kindlier' mountain. School parties, and other parties, organised by youth organisations involving juvenile participants, invariably choose Beinn Alligin for a ridge walk. In other words, I believe that the majority of hill walkers tend to view Beinn Alligin rather too lightly. Statistics of the mountain fatalities over the twenty-one-year spell of my time at Torridon do show, in bald figures, that, of the nine unfortunates involved, five died on Beinn Alligin and four on Liathach. I am not going to make any direct comparison myself; I am simply going to say that one cannot view any mountain in Scotland lightly, all merit respect.

On Beinn Alligin perhaps the most interesting, nay, fascinating feature (except to the ridge walkers) is Coire Toll a' Mhadaidhe, referred to earlier. There are various 'memorials' dotted throughout the Highlands testifying to the presence of the wolf in our native fauna at one time. This coire is, to me, the most striking of all of them. Various dates are given for the killing of the last wolf in various parts of the Highlands. There is no definite date for the last wolf killed on Beinn Alligin, simply a local belief that it was killed here in Coire Toll a' Mhadaidhe. The more or less accepted record for the demise of the last wolf in Scotland is 1743 in the River Findhorn area, killed by a MacQueen of Polochaig. I find it entirely credible that Coire Toll a' Mhadaidhe once afforded sanctuary for the wolf, for so deep and intricate is its network of cracks, clefts and concealed rock passageways that one can well visualise it hiding a human raiding party, in the long ago days of clan warfare, never mind a wolf or two. It is interesting to realise that the clansmen who fought at Culloden, in 1746, may well have had among their numbers some who were familiar with the wolf, yet had no knowledge whatever of a brown, cuddly animal called 'rabbit', which, as a matter of record, did not reach the Wester Ross-shire area of the Highlands until 1850, nearly one hundred years on from the demise of the last wolf.

I found that the entire Beinn Alligin ridge took less time than the ridge of Liathach and that of the Horns, the only bit which needed some rock-scrambling skill (i.e. handholds as well as footholds), was shorter and simpler than the Pinnacles of Liathach. However the Horns should never be taken too lightly for there is a quite spectacular sheer drop in their north side.

On the Beinn Alligin guided walks one had to exercise discrimination as to who was capable of doing the rock scrambling section; finding out that someone had a bad head for heights had to be ascertained before one was in the middle of an exposed section.

On Liathach one could always be assured of seeing plenty of deer. On Beinn Alligin one saw less deer but, on the other hand, it was odds-on that you would see a golden eagle and the almost universal cry from visitors was the desire to see either red deer or eagle and if possible both. The eagles had perching roosts, or resting places, on the towering cliffs on each side of Coire Toll a' Mhadaidhe, different ones being used according to the wind, for shelter, while sitting, immobile as a graven image, for hours at a time, so as to digest a full crop of food. The finest sighting I ever did have of a golden eagle was from the Horns of Beinn Alligin when it rose from directly below us, not twenty yards away, and swept around the Horns towards Beinn Dearg. It was truly magnificent, looking down at the seven-foot wingspan just below, with all the buffs and pale browns, the gold of the head and neck clearly visible, instead of the black silhouette which is usually all one sees of an eagle looked at from below, against the sky.

Amusing incidents, perhaps telling incidents, occurred with such a diversity of walkers. A very nice middle-aged couple from Holland turned up on one occasion to walk on Beinn Alligin. The husband was carrying an enormous rucsack and I wondered if I should ask him if he required to carry such an apparent weight to the hill. Luckily the problem was solved for me; he checked the contents of his rucsack as we stood in the car park. They included a two-gallon polythene container of water. Having had a drink of hill water just whenever I desired I was dumbfounded. I told my walkers that there was no need to carry water in the Highlands; in incredulous stare rewarded me. 'Nowhere in Holland can we drink from a stream' he told me, 'we have to carry water.' I had quite a time reassuring him that it was indeed safe to drink from our burns. The water from burns on Liathach and Beinn Alligin was the purest and best that I have ever tasted, ice cold, at any time of the year.

On that same walk we saw multitudes of tiny, sparkling, green frogs. Again this caused wonder in my walkers from Holland. 'We have no frogs left in Holland,' I was told. Was this Torridon area, in view of this, a disadvantaged area, or was Holland? It, obviously, depends on your point of view.

While on the subject of frogs, I remember an incident which had me believing I had had a major accident to my party, which was strung out behind me. Walking along, on a circuit of the lower slopes of a coire, I was suddenly halted, petrified, by an ear-splitting scream, terror implicit in its shrill notes. I spun around, expecting to have, at the very best, a broken leg. No such thing! A teenage girl, half-hysterical, was sobbing, one hand to her mouth, the other outstretched, pointing. Pointing at what? I went across expecting to see an adder and ready to reassure her. Sobbing still, the girl managed to get out, 'It's a frog! I can't stand frogs,' the last few words in a rising falsetto shriek.

There was another day on Beinn Alligin, high up in the screes, when it was my behaviour that was the cause for conjecture by onlooking watchers. Just as our party was nearing the top of Tom na Gruagach I spotted, ahead of us, ptarmigan superbly camouflaged as always, feathered, counterfeit, grey 'rocks', amid the genuine grey rocks. They were five of them, obviously a brood of the year. I pointed them out to my walkers as the ptarmigan scattered, while each individual grey-flecked bird weaved a devious way through the rocks. Another hillwalking party had arrived at the top only minutes ahead of us and had seen no ptarmigan. Nor did they see them, as I, camera in hand, crouched over, sometimes crawling, went in pursuit of a photograph of the elusive grey birds of the mountain. Various theories were advanced for my crawling about in pursuit of 'nothing'. 'Exhaustion, poor fellow!' 'Effects of rarified air on the tops.' 'Just one of those daft geologist chappies.' I am happy to say that I got a photograph.

There were always incidents, or people, one remembered. I distinctly remember the two Israelis, of about twenty years of age, whom I had out on a guided walk. We talked of snakes and I told them of how many people in Britain were scared of snakes. The Israeli girl, matter of factly remarked quietly 'The only living things I am frightened of are other people.' The lad, when I remarked on the huge rucsack he was carrying, said 'I was on army training just before I came on holiday. We carried very heavy packs and our automatic rifles and ammunition, much heavier than these packs we are carrying up the mountain. Of course, you do not need to carry automatic rifles, in Scotland, to protect yourselves.' Simple statements, yet so very eloquent!

Then there was the family from Dublin, Mr and Mrs O'Connor, who travelled all the way from there, to Torridon, and took bed and breakfast in Inveralligin for one night. This journey was undertaken, with their nine-year-old son, for the sole purpose of giving him a chance of seeing a golden eagle. He had been doing a project on the golden eagle at his school in Ireland and had got absolutely obsessed by the subject. Someone gave the parents my phone number and they phoned and asked me what were the chances. I said that I thought we could manage if the laddie was a reasonable walker. They journeyed over on that slim assurance, stayed the night at Inveralligin and met me next morning. I knew of an eyrie where the eaglet was nearly ready to fly on its first flight from the eyrie and off we went to look at this nest. The laddie, buoyed up with effervescing excitement, walked well; his parents were obviously not really interested in eagles and they did not walk too well. After all, the hill ground was rather different from the streets of Dublin. All honour to them; they had taken a trip into the wilds of Torridon, where they obviously felt ill at ease, simply to let their boy see an eagle. How many parents would have done this? We eventually got to where I could let the laddie have a good look at the well-grown eaglet on its eyrie; the pure joy on his face was well worth all the effort. I firmly believe that our young folk should be helped, in every way possible, to see and learn about our native wildlife, in their natural surrounds, with all the effort and difficulty involved. 'My' Irish family began their long journey back to Dublin just as soon as I got them back to the roadside. I hoped the laddie's interest in the golden eagle would flourish and expand over the years ahead of him.

There is an enjoyment, or exhilaration if you like, to be won from wild weather on the hill also. I had a party of folk out on one such day when squalls of rain blew intermittently before a gale-force westerly wind. We went into one of the coires, for the ridges were quite out of the question, indeed had I not known my walkers were of tried ability I would not have gone hillwalking, even into the coire we went to that day. The wind was fierce and behind us on our way out; we had a damp lunch cowering out of the wind, while watching the grey waves scurry, white-capped, across a hill lochan. Lunch was brief, and the homeward way was toilsome into the wind, the wild wind which now sent us reeling off balance at times. Yet we all enjoyed it, this struggle with the elements, for we knew we were equipped and experienced enough to be able to cope with them.

An unforgettable and certainly not an enjoyable memory of Beinn Alligin was of a heather fire we had there, on an April day after we had had a dry spell of six weeks. Aided by a friend, Tom Wallace, I worked at this from 10.30 am until 5.30 pm that evening, without a break for the fire was not going to pause for this. We fought and fought at it all day, with birch brooms, gradually forcing its raging, waist-high flames towards a long rock outcrop which would check its advance. Time after time we despaired of achieving this; time after time we dashed back in and renewed our desperate efforts, battering at the flames. One lost ground, blaspheming silently, or at times not so silently, when the wind gusted strongly and fanned the fire into renewed life. When the wind lulled we worked like demons and gained ground perceptibly, but every now and then an audible sharp crackle behind had us turning and racing back to put out, anew, an area which had re-kindled. Nearly every burn, which otherwise could have been depended on to check the fire, was bone dry after the lengthy drought. The sphagnum moss had a dry top crust, over which the flames raced; the dry, grey-green cushion moss was, on the other hand, like tinder and very hard to subdue, smouldering sullenly, to fan into red flames long after you believed that you had put it out. So dehydrated did we become that we drank voraciously at any ash-strewed, stagnant pool of water we were able to find. When finally we had got the long, ascending line of flaming heather under control we were pretty well exhausted. I had fought a few hill fires in my day but that long battle was the worst ever, in incredibly dry conditions on the hill.

Reclining, worn out, on the hillside we watched for some time to see that the fire did not re-kindle. So dry was everything that the underlying peat had been glowing red at times. We were conscious of aching limbs and an unwillingness even to rise to our feet to go home. My face was painfully scorched, my eyes prickled as if with a thousand needles; Tom told me that I had no eyelashes left. I had holes burnt in my stockings and in my breeches, and red burns patched my legs. My hands were so blackened and blistered that the feel of the steering wheel driving home was almost unbearably painful.

Not an enjoyable memory but certainly an endurable one.

4. Mountain Rescue

When I left for Torridon I had spent more than twenty years almost daily walking the hills which rise above the Great Glen of Inverness-shire, at all seasons and in all weathers, mostly on my own. Never had I been conscious of the need for rescue – if you could not look after yourself on the hill then you should stay at home. That, with probably ill-placed confidence, was my philosophy. The rolling ridges and peaty flats of the Western Monadhliaths were not thought of as 'dangerous' hills, nor were they sufficiently impressive, scenically, to attract many hillwalkers.

The concept of mountain rescue was quite novel to me therefore prior to the move to Torridon. Liathach, that truly stupendous mountain, dominating all below it, with its long narrow summit ridge of splintered rock fangs, its girdles of sheer sandstone ramparts, its wide sweeps of shattered rock screes, changed all that. With neighbours all around of more than 3000 feet in height, challenging for interest, and numerous hillwalkers ready and eager to accept this challenge, the majority of these from urban areas, a mountain rescue facility was an obvious necessity. There was already a mountain rescue team in the area, led by Charlie Rose, composed entirely of volunteers, and this team I gladly joined.

Any misguided notion I had had about the glamour of mountain rescue work speedily vanished after my very first outing with the team. Mountain rescue work I found, was tough, very arduous, mentally and physically testing, frightening on occasions, and one was always emotionally involved. Try as I did to avoid it, I always tended, in the case of a fatality, to identify with the casualty. Here was someone, like myself, to whom the hills were the very breath of life, and now that life was gone. Emotional stress, perhaps over-emotional stress, drained me at such times; one felt unwilling to introduce to the hill, that impersonally cruel hill which had struck once again, yet more walkers.

Nor did mountain rescue work run to a tidy or predictable pattern. Most often, at the end of a busy day, as you had your feet up at home, relaxing, too weary to even read a book, the call-out would come. One cursed a little, flexed stiff legs, struggled anew into hill gear and got cracking to join the team.

Feelings were always a bit mixed when you were directly involved in a rescue; anxiety always as to whether the team could locate the casualty and be in time to save life. Tremendous relief when you did succeed in this. Frustration when the person for whom you had searched the hill, for perhaps the best part of a dark, cold night, arrived back, safe and well, after you had set out to succour them. Exasperation, as when an indomitable but misguided old lady of seventy-five had embarked, alone, on a strenuous hillwalk upon an October day of utter savagery, of horizontal flint-like sleet, driven before a gale force wind and – scarcely surprising – had gone missing. Or when, equally foolishly, a party of three hillwalkers had set out, in thick mist which never did clear that day, to do a ridge walk. Only two returned. Comedy also, thankfully, as when a couple of hill-walkers came one evening to beg for the use of our mountain rescue stretcher, telling us that their dog had collapsed on the hill. Astonished, thinking in terms of a terrier-sized dog, I asked why they could not have simply carried it down between them. A sheepish look ensued. 'We're sorry – its much too large – it weighs around thirteen stone.' Amusement also when a walker, a member of a well-known mountaineering club, broke a leg, on a December afternoon, on the tourist path. His over- riding concern proved to be that his club must not hear of his 'ridiculous' mishap.

Admiration for uncomplaining endurance as when one helped to improvise a splint on a nastily-broken ankle while the courageous casualty lay on sharp-edged rocks, or when a youngster, similarly disabled, had to be winched up, from so dangerously vertical a hillside that it was impossible for the rescue helicopter to land, to the waiting helicopter, high above. This particular rescue technique is awe-inspiring to watch; I should explain that the winch cable upon which one is suspended while being immensely strong, appears thread-like and utterly frail in these circumstances.

Admiration, above all, unstinted and appreciative, for the sheer skill and raw courage of all the personnel of the mountain rescue helicopters. They turned out to help in all kinds of inclement weather, impossible weather one felt at times, and manoeuvred, with incredible precision, into almost impossibly hazardous positions, to achieve a rescue.

I shall always remember my first involvement in mountain rescue work. It came, typically, one evening after most of us had put in our day's work, and involved going to search for a hillwalker, known to be dead because of a bad fall, in a remote rocky coire. There is no mental stimulus in such a rescue, no 'lift', in that one is not buoyed up by the hope of saving someone unfortunate enough to have been injured on the hill. One feels sympathy, compassion, a feeling that it might well have happened to you, but it is a sad duty, rather than a rescue attempt illuminated with hope. We assembled about 8.30pm, with still plenty of daylight left and we had had an approximate situation of the casualty from a companion who had come down to raise the alarm. Nevertheless we only just beat the darkness in eventually locating the casualty in the rock-strewn coire. I found it hard to accept that, only that morning this chap had set out to enjoy a day on the hills, now becoming familiar to me, and was now lying, so still and silent, in the quiet of the onset of dusk on the surrounding hills. The return journey, sombre in spirit, was fittingly in darkness. Thankfully the sheer physical effort and concentration involved, in carrying a heavy man, while steering a necessarily erratic course through rocky and dangerous terrain, gave one little time to brood. Our team split into three carrying groups which ensured that each group had five minutes of carrying duty and then ten minutes relief. Predictably, the five minutes of carrying duty seemed arduously endless, slipping, slithering, clattering your shins on rocks, all the time intent on maintaining the balance of the stretcher. The ten minutes off-duty, by contrast, went like lightning. The night was one of velvet-blue, brilliantly lit by the stars, a lovely night but for our mission. By 1am we had arrived, leg weary and arm weary, at the road above Loch Torridon where we had left our vehicles. Glamour had gone for ever; grim reality had replaced it, in my outlook on mountain rescues.

There were rescues, for one reason or another, which remained vivid in one's memory. An account of some of these will, I'm sure, give my readers a better understanding of what mountain rescue in Torridon involved.

Mountain rescue work throughout the entire Highland area was brought to a pitch of much increased efficiency in the early 1970s, with the facility of being able to enlist the aid of the Royal Air Force mountain rescue helicopters and their supremely skilled and dedicated crews. Many, many hours of strenuous and sometimes frustrating work were saved. An aerial search by helicopter was much more likely to be successful in such broken, rugged terrain as ours. Perhaps most valuable of all however was the ability of the helicopter to airlift injured hillwalkers directly from the actual scene of the accident to the nearest hospital. In the case of Torridon, our nearest casualty hospital was Raigmore, in Inverness, approximately two hours by road. This same journey took twelve minutes by helicopter. Add to this two hour journey by road the amount of time needed to carry a badly injured casualty over incredibly rugged terrain and you could have a very considerable total time-saving of four to five hours. From this time also, in the majority of cases, the casualties, picked up and airlifted to hospital directly off the hill, were warmly tucked up in hospital, after receiving treatment, long before the rescue team, weary, but contented at a good job done, were home at their own firesides.

Early in 1974 our Torridon rescue team had its first full participation in a joint helicopter/rescue team exercise, in a call-out to aid an injured hillwalker. Inevitably, I had been out on the hill all day and arrived home to receive the rescue call-out. I had time for a hasty cup of tea and then I was off again to rendezvous with the team. The injured walker, we had been told, was lying up on the summit ridge of Beinn Eighe, with a badly broken leg. She had fortunately been one of an organised party. On their way home, on the ridge of Beinn Eighe, which is an absolute jumble of chaotically scattered, sharp-edged quartzite blocks, she had put a booted foot between two of these blocks, tripped, fallen forward, and snapped the firmly-trapped leg just above her ankle. The accident was a classic instance of a fact I came to accept which was that mountain accidents most often happened on the way back from the hill. Factors which we believed led to this were tiredness, after a strenuous day, a certain lack of concentration due to this, and perhaps a certain eagerness to hurry a little, to get back and get that welcome cup of tea, or pint of beer. The sum of these led to a proneness to accident, particularly in difficult terrain.

I arrived late at the rendezvous, at the base of Beinn Eighe. Most of the team had gone ahead leaving the so-called lightweight mountain rescue stretcher there, with a message asking that 'an active team member' bring it up. Since I was known as an 'active' hill man I took it that the message had been left for me. Wriggling my shoulders into the carrying straps I must confess I regretted the cup of tea that had kept me late. Bent over more than a trifle by the weight of the bulky stretcher I nevertheless caught up with some of the team who had gone ahead of me and, near to the ridge, was relieved of the weight of the stretcher, none too soon either, by a husky colleague. By then I had caught up with the surgeon who, luckily, had also been hillwalking in the area and who had volunteered to come up and help aid the casualty.

It was a day which had been bright and sunny, towards the end of May, but with a strong north wind and this, blowing strong and cold on the tops, had almost certainly contributed to the accident. Indeed, on these very tops, we had had an occasional shower of fresh snow that day, which added the complication of slippery going. The surgeon and I arrived at the casualty together, where she lay, plucky lady that she was, under a perfect mound of jackets and other clothing, divested by her companions, in an attempt to keep her warm. Most of these companions had now gone down, secure in the knowledge that help was at hand.

Lying on that exposed ridge, at 3000 feet altitude, the casualty needed that mound of clothing for she had been lying there, virtually motionless, head downhill as she had fallen, for five hours, by the time we arrived. With the precept, do not move an injured party until medical aid arrives obviously engraved on their mental tablets, her party had not moved her at all. Laudable, I felt, but a trifle misguided under these circumstances. They had done everything else possible to make her comfortable however, and, stout-hearted lady that she was, she complained not at all. Carefully, gingerly, we manoeuvred her onto the stretcher, under the direction of the surgeon. He administered some morphine from our mountain rescue medical kit and, since I was the nearest to him, he turned to me and asked if I was squeamish at the sight of blood. I replied that I was not, and he then asked me to assist him in treating the injury. Though I had said that I was not squeamish the actual close-up sight of the jagged end of broken bone sticking out through the flesh did give me a qualm, and ever more respect for the fortitude of the casualty. First giving the morphine time to take effect we then, as gently as possible, removed boot and stocking. Next stage was to manipulate the leg into position for setting it, temporarily, until the casualty got to hospital. Under the surgeon's explicit instructions I then maintained traction on the leg, (simply a steady pull on the leg), while he improvised with a temporary splint.

By this time occasional showers of snow were beginning to sweep the tops again, and it was decidedly chilly, the body heat generated in the climb to the ridge now dissipated. We had been told to expect a rescue helicopter to arrive approximately at 9pm and, inexperienced in the calibre of these airborne rescuers, we began to wonder about the feasibility of this, as the snow showers thickened. Our next priority however was to carry the injured walker, now strapped securely on the stretcher, to where the helicopter could probably make a landing to airlift stretcher and occupant. A very careful slow progress ensued, picking a way first along the slippery sharp-angled blocks of quartzite rock. Down then, via a steep, uneven grassy slope, slippery with fresh snow, towards a tiny 'flat' where we thought the helicopter, a Wessex, could land. As if in malign intent, to render things even more hazardous, a heavy blizzard of white, whirling flakes now enveloped us and in seconds we were all plastered white, the slippery slope now even worse to negotiate. Very thankfully we reached our objective without mishap and even more thankfully we heard, but could not see because of the blizzard, the racketting approach of the helicopter.

Dismay replaced relief as we realised that we could not be seen either from the helicopter and this dismay deepened as the racketting noise swept past and beyond us. Twice this happened as we stood, white-shrouded, waiting, and hoping. As suddenly as it had begun the blizzard thinned, we saw a glimpse of sky, the racketting noise of the helicopter increased and as if by magic the bright yellow shape of the RAF rescue Wessex helicopter loomed through thinning snow. The wind strength was now very high but the pilot brought it in, hovered momentarily, as if testing the wind speed, and then, so delicately as to seem unreal, he brought it down and, making contact only with two offside landing wheels, held it balanced there, defying the wind. The door in the fuselage opened and two leather-jacketted, helmeted figures dropped out and ran towards us, gesturing urgently at us to keep our heads low, to avoid decapitation by the whirling blades of the rotor. Crouched low, we quickly brought the stretcher forwards, slid it into the helicopter and then they were vanishing into the snow mist again. The total time spent in 'loading' the balancing helicopter was just under three minutes.

Vastly impressed, appreciative and jubilant, our return down the hill, that May night, was enjoyable. Our casualty was in Raigmore hospital by 9.15pm, warm and safe. We arrived home around 10.30pm a modest glow of achievement warming us.

In mountain rescue work one was often called upon to exercise one's own judgement as to whether it was necessary to call out the team or not. In one such case I was able to do the entire rescue alone.

I had just arrived back at our countryside information centre from a hill walk, in summer, when a distraught chap burst in, gasping incoherently 'Can you help me, please?' He told me that he had had to leave his wife, two young children and their terrier stranded on the vertically steep, heather and rock-strewn side of a deep gully which rose about 300 feet above a rocky burn. The family had walked up alongside the burn and rather than go the same way back, had elected to climb straight up the side of this gully. They had over-estimated their joint capabilities; two thirds up the interminable, now direful, side, they became stuck. The father had secured the children to a birch sapling, the mother and the terrier beside them, while, in utter desperation, he steeled himself to find a way up the terrifying slope. From his description I knew where they had got stuck and away I went with the already tired father. When we arrived I left him resting above (he really needed to) while I worked my way down and located the family, by now tearful and distressed, but otherwise fine. I was then able to carry, pickaback style, each child, one at a time, up the steep side. We progressed carefully crabwise, gradually slanting upwards. Both children were quite composed by now; this was adventure! I'm afraid that I could not carry their mother out but she, knowing her children were safe, gritted her teeth and slowly, step by step, we climbed out, the terrier secure in my rucsack. The joyful reunion was reward in itself. To the children, at least, it was now an adventure to savour.

The team experienced a memorable rescue one sultry evening in August, memorable because of its blinding, grey swarms of midges and because we spent the night at an altitude of 3000 feet, perched out on the Pinnacles ridge of Liathach. Three hillwalkers, a father and two sons, had been making their way down from this ridge when the thirteen-year-old younger son had slipped and fallen into one of the rock gullies which seam the south face of Liathach, gullies which are near vertical in their steepness. A lucky laddie, he lived to tell the tale; he sustained only a broken ankle in his fall. His father, scrambling down into the gully, stopped with him while the other lad, seventeen years old, set off down for help. In his necessary zigzag progress, threading his way among the chaotic jumble of rocks, screes and precipices of that steep face, he became, quite understandably, a bit disorientated, but eventually reached the Glen Torridon road to raise the alarm.

It was 8.30pm by the time the first of us had reached the lonely house, Glen cottage, which was to act as our base for the rescue. The evening was warm and dry, ideal, but the midges also considered it ideal and were swarming in voracious hordes. The laddie who had come down was very tired but seemed very composed and assured us that his father and brother were in that sheer deep gully, at times used as a direct line of assault on Liathach, which runs up directly behind Glen cottage. With this very positive identification of the gully where the mishap had occurred the subsequent rescue seemed simple, if arduous; we would go straight up, locate the two walkers and bring them down. Because of this, and the sultry warmth of the evening, most of us left our heavier clothing down at base. Since I knew the hill well I was asked to go ahead to try and locate the walkers, while the rest of the team assembled. I carried the stretcher for that first leg. The face was quite uncompromisingly steep so that periodic rests were forced upon me, voracious midges seizing on these 'restful' interludes to ensure that I did not rest long. Rock outcrops barred the ascent at intervals and were hazardous to thread through, with the unwieldy stretcher seemingly imbued with devilish intent to pull me backwards off the rock as I inched upwards.

Eventually I reached where I thought was the approximate position pointed out to us by the elder son. There I halted, very thankfully, and, since I could only see part way into the shadowed gully, began to whistle and shout to establish contact with the boy and his father. Only echoes answered me; nonplussed, I climbed a little higher and tried again; again no reply. The forerunners of the main team were now approaching below me; we were now around three-quarters of the way up to the top and still had no location of our walkers. Anxiety began to rise in me; this rescue was not, after all, going to be so simple. Sticky with sweat and encrusted with dead midges, I sat down to await the team. Puzzled anxiety was now partly dissolved by the arrival of Charlie Rose who had just had a message from base, via our walkie talkie, to tell us that the laddie, who had been so sure of the position of his father and brother when we had set out, had now altered his mind, radically. He now thought that they might be in a gully at least half a mile westward of the one we were now investigating. Comments and opinions were forcibly expressed, sadly, quite unprintable.

We were now in a quandary. There was not a lot of daylight left. We could not safely cross that deep gash in the hillside, in order to search the gullies to westward, unless we descended almost to our starting point. Alternatively we could continue up, since we could also cross at the top of the gully, and we had already put an awful lot of effort into climbing near to the top. We should have to stay the night out on the ridge, but the night was warm and dry and we could resume our search at daybreak, thus saving both time and effort. In the now fast-fading light we slowly, and in some cases painfully, resumed the steep climb to the top. Darkness had come down before the stragglers of our team, climbing necessarily in single file, achieved the ridge. Those last arrivals had to home in on our shouted directions, so dark had it become. On the narrow path below the ridge we spent the night, strung out like swallows on telephone wires, crouched, knees up to our chins, unable to get really comfortable because of the dangerous steepness of that face. Only the indistinct lightness of the night sky relieved the darkness. Above us, just discernible against this sky, there reared the jagged fangs of the dreaded Pinnacles section of the Liathach ridge. Perhaps more descriptive was the Gaelic name, Am Fasarinen, which means the Fangs, a far from over-imaginative description of these ages-old, elements-splintered, gale-sculptured, shattered rocks. This was a ridge which inspired awe, even in bright sunlight; sitting up there, each one isolated by the oppressive darkness, and one readily conjured up the spectres and monsters of Highland superstition.

Having said that, it remains an experience I should have hated to miss.

Few of our party slept at all that night, if only because of the unspoken fear that we might roll off our precarious perch and away down the steep face which yawned, unseen below us. One of our party of twenty, lucky or unimaginative, certainly managed some sleep, by the sound of his snoring. I do not know to this day who it was for the darkness was so palpable that one could not see one's neighbours. Now and again a bag of dried raisins was passed from unseen hand to unseen hand, groped for clumsily but accepted gratefully. Once a large packet of biscuits rustled its way, invisibly but audibly, along the line – no one was really hungry but it helped to pass the time. Occasionally the flare of a match disclosed a chap trying to derive solace from his pipe – at times a muted conversation started and quickly faded. Towards the first eagerly watched for indication of dawn a few light raindrops fell and in the very literal chill of pre-dawn a bout of uncontrollable shivering and audible teeth chattering became apparent. A few shivery, scattered notes, dropped reluctantly it seemed, by a rock-perched ring ousel, I heard at 3.30am. By 4am indeterminate shadows of grey were assuming human shapes, crouched below the toothed summit ridge. There came a stirring into life of these grey shapes, a rustling of jackets, a clearing of throats, a click-clack as booted feet were struck together to warm cold feet and the thump of arms thrashed against chests. It was another half hour however, before, breakfast-less, we could get moving, reasonably safely, along that precarious ridge. As we passed the head of each gully we peered down as far as we could see, shouted and whistled, then moved on along the ridge.

About an hour later, we had got to the end of the Pinnacles section, and we begun to descend a bit, preparatory to searching each lower part of the deep gullies as we came to them. It was full daylight by now and the sun was already warming us. With the full daylight too we had been able to resume contact with our base. The first gully we drew blank at; we were approaching the next one when our walkie talkie crackled into life to tell us that base could now see us and that, around 200 feet higher than the line we were now on, in the gully we were now approaching, a figure in an orange anorak had appeared, waving his arms as if trying to signal. Relief and jubilation illuminated our entire party; this could only be the father of the injured lad. I was deputed to forge ahead once more and ten minutes later, after a faint reply had come to my shout, I was scrambling into the deep gully and reassuring father and son that rescue, at long last, had arrived. Back I scrambled to guide the team to the lip of the gully. From there half a dozen of us went in and, first attending to the broken ankle (badly swollen but with no end of broken bone protruding) helping the laddie out of the gully. This was too narrow and difficult of access to be feasible for our stretcher. Rigging a safety rope the laddie was able to hop out, courageously, on his good foot, while we supported him until we were out onto the sweep of the hillside. Base had been radioed that we had found the pair and had, in turn, summoned the aid of a rescue helicopter. We were all, rescued and rescuers alike, happy to rest, relaxed and relieved, on the steep face, enjoying the warmth of the sun, the glen road a grey winding ribbon far, far below.

As always the racketting thrum of the helicopter first alerted us, then, moments later, we saw the familiar, welcome, bright yellow whirly bird come in view down the glen. It racketted by, almost on a level with us, continued past until obviously out over Loch Torridon, turned in a wide sweep and flew back towards us. We were now waiting to find where we would have to transport the injured boy on our stretcher to where the helicopter could land. On that steep seventy degree slope it was going to be difficult. Judge of our relief when the radio told us that it was to be a winching-up operation; the helicopter was not even going to attempt a landing, nor did they want our stretcher; they had their own stretcher. Nothing remained but to clear the pick-up zone of all un-needed bodies, leaving only three of us to assist there, with the injured lad. The thrum-thrum of the helicopter grew louder and then it was hovering above us like a giant yellow dragonfly, the downdraught flattening us protectively to the hillside, feeling indeed as if we were to be plucked out into space from this.

The scanty grass stems on the hillside bowed in homage also, a bulky figure emerged backwards from the opened side, to dangle on an absurdly thin line, like an outsize spider. To our fascinated gaze it appeared to be astride a short wooden board. Down the apparition came, rapidly, twisting and twirling, but with controlled precision, until we could reach up and guide his feet down. The bulky spider-man grinned cheerfully at us and unhitched his wooden bench, before signalling the helicopter away to a more comfortable distance. The winchman unfolded the 'wooden bench' which became transformed into a collapsible stretcher. Under his direction we gently strapped the lad onto it; we had warmed to him for his courage even on our short experience with him. Living up to this, he was bravely concealing his very natural apprehension at being whisked into the air, the glen yawning below. Back came the helicopter and again we clung to the hillside; the winchman attached the dangling cable to the stretcher and checked, then double checked, the fastenings. Thumbs up; a quiet 'Thank you' from our rescued laddie, and up he was whisked, the winchman beside him, reassuringly. A momentary pause opposite the open side of the helicopter then they were guided in and were off. That air of euphoria, common to all successful rescues, enfolded us as we set off for the Glen cottage, tiny, far below us. It was exactly 7am; we were goin to be late for breakfast.

Perhaps fortunately, between the more lengthy, more dramatic rescues, there were rescues with elements of light relief, as when we stretchered back the Irish wolfhound, referred to earlier in the chapter. This rescue was quite obviously not in the least bit funny to the owners of the distressed dog who were extremely concerned for the plight it was in. They had been given the stretcher on the afternoon they had asked for it and had toiled away back out to the Horns of Ben Alligin where the dog lay. There they found that they could not manage unaided; the dog was too heavy and the terrain too relentlessly steep and rocky. Consequently we were appealed to once again on the following morning and this time we scraped together sufficient volunteers from our team to go out to rescue the wolf-hound, guided and aided, by the owners' party.

It proved quite an undertaking, involving for one thing, lowering the dog, by now warmly wrapped up and securely strapped onto the stretcher, down the more precipitous steeps. I believe it was now abundantly, if belatedly, clear to the anxious owners just how ridiculous it had been to attempt to take a large, heavy dog, on a sultry day in October, on such a potentially hazardous ridge as the Horns of Ben Alligin. Their poor dog was quite comatose still; I believe that a combination of utter exhaustion and dehydration had caused its collapse, on the preceding day. Don't get me wrong; not one of our team, who had volunteered to aid in the rescue, grudged the time and effort involved in eventually getting the wolfhound safely off the hill, but the Horns of Ben Alligin was not a place any of us would have taken an Irish wolfhound. The said Irish wolfhound was obviously revived a bit by the time we had got him back to where the owners' car was parked. I shall be accused of exaggeration when I venture to suggest that there was even a hint of an Irish twinkle in his now more alert Irish eyes.

All of us aided in the gentle transfer of the wolfhound from its warm bed on the stretcher to an equally warm bed of blankets in the car. Still stretched flat out, but safe, it was driven away. We, for our part, made our separate ways home, grateful thanks ringing in our ears, and, possibly lingering longer, the memory of the warm, brown, uncomplaining eyes of our Irish wolfhound. A quite unforgettable rescue for obvious reasons.

On a gloriously sunlit day in July with thin, sun-filtered gossamer-like veils of mist alternately concealing, then revealing, serrated mountain tops, etched against a deep blue sky, I arrived home in the evening to be told that there was a mountain rescue alert on. At once I contacted the mountain rescue post to be told that this was indeed so and that some of us were to be flown out in an RAF mountain rescue helicopter, which was due to arrive at any moment, in an attempt to get to the scene of the accident quickly. The alarm had been raised by another ridge walker who had been on the Pinnacles ridge of Liathach, nearby, when the accident had occurred. One of a pair of two ridge walkers on the Pinnacles section had lost her footing and had fallen.

I changed my stockings and boots, chucked some odds and ends of food into my rucsac and hurried up to the rescue post where some of the team were assembling. My hurry proved justified for hardly had I arrived when we heard the now familiar discordant clattering which bespoke the arrival of the Sea King helicopter. It landed on the nearest piece of flat ground to the rescue post and consultation with the crew was kept to a minimum – circumstances, approximate locality. Then some of us piled aboard, ducking low in the fierce whirlwind engendered by the ceaseless whirling of the rotor arms. Time was obviously of great importance; those of us who knew well the rock fangs of that ridge, its extreme narrowness and the awesome drops to each side, knew that anyone who had fallen there was likely to have sustained severe injury. Safety belts buckled, earmuffs on and we were lifting off, as smoothly as is possible in a machine of which the very framework was shuddering and juddering under the racketting torment of the rotor blades.

We tilted around the end of Liathach with ridiculous ease, compared to the arduous and time-consuming footwork we would otherwise have to have undertaken. My view was restricted to what I could see from a small square window and was not markedly different from what I had seen, scores of times, from the 3500 feet summit of Liathach. The same tiny lochans bejewelling the lower slopes, girdled by walls of containing Torridonian sandstone; the same long, thin river, stretching its lengthy tentacle around the base of Liathach; the same red-brown dots, against the greens of the upper coires, which were red deer, at peace before the helicopter set them running.

Half a dozen times our helicopter edged its crabwise progress parallel to the cirque of black rock crags which dropped in a sheer, gully-seamed, semicircle from that rock-fanged Pinnacles ridge to the screes far below, each time at a different level, each time with straining eyes striving to spot the accident victim. The ceaseless, relentless clangour was redoubled as it was echoed back from the rock walls which we were searching with the helicoper skilfully piloted as close to this rock wall as was humanly and mechanically possible.

I saw the rucsack first, high up in a narrow deep-shadowed gully, black edged and rock fretted. It had obviously been torn off in the fall. I followed the dark gully downwards with my eyes, mentally shrinking from what they must see next. Near the bottom of the gully, sprawled out, face down to the rocks, absolutely lifeless-looking, lay a body.

This confirmation of what, I believe, we had all of us feared, was nevertheless not pleasant. Indeed, on such a lovely evening, it seemed illogical that tragedy could lie so close. It should rather have been pouring rain, a grey, sodden, drenching evening, with those stark crags weeping in mourning.

Since the approach on foot would have been time-consuming and potentially dangerous, the helicopter crew decided to put the winch man down as near to the casualty as possible. If this attempt failed, we would then be dropped, as near as possible again, to stretcher her to where the helicopter could lift her out. First, however, in order to lighten the helicopter and thus render this more manoeuvrable under testing conditions our rescue party was landed on a rock studded 'flat', below in the coire. This achieved we watched the now lightened helicopter begin to edge in, even closer than before, to these daunting walls. Poised as directly above the bottom of the gully as was possible, hanging in the air like a huge, hovering, yellow vulture, the helicopter then disgorged its winchman, a-dangle on the thread-like cable, down, down, down towards the sprawled out casualty. Freed by the winchman's successful landing, the helicopter circled back out to hold a watching brief. We all watched the tiny, distant, courageous figure of the winchman achieve his objective. He waited at the crumpled body while the helicopter circled back, paying out winch cable as it hovered in position again, above the winchman. At the end of the cable this time was a sling, dangling at the end of the cable, which was to be used in winching up casualty and winchman. This had to be grasped by the grounded winchman as it swung in close above him as he perched, precariously, on that steep rocky face, an exercise in fine, synchronised, judgement between pilot above and winchman below. The initial attempt failed, the sling swinging to and fro just out of reach of the winchman's grasp. Out the helicopter circled once more then again crept close in to that dangerous wall of rock and, this time the winchman was able to grasp the sling. I had cherished a faint and quite illogical hope that there might linger yet some salvageable breath of life in that sprawled-out body. Seeing the body dangling, utterly and irrevocably lifeless, with the winchman accompanying it, as it was winched up, rotating slowly, into the hovering helicopter, extinguished even this tiny glimmer of hope.

It was a subdued lot of would-be rescuers which was picked up a few moments later by the helicopter, its noise alone unabated and unabashed. We had all gone out to attempt to rescue a life on a halcyon evening when the harsh finality of a sudden, violent death on the mountain, seemed impossible to accept. The beauty of that evening rendered the tragedy the more poignant. We were now coming back with the sombre reality, shrouded and still, in the helicopter with us. I heard in the aftermath that the victim, in her late twenties, had spent her last fourteen years in climbing and hillwalking in this country and abroad. On that day, on the Pinnacles section, leading the way down a kind of knife-edged rock staircase, the sheer drop to one side, her companion following, she had suddenly slipped, sideways and outwards. Rivetted, helpless, utterly incredulous, her companion looked on while she fell, cannoned into a rocky projection and in much, much less time than it takes to describe, bounced off this, to plummet into space and out of sight. One moment glorying in her mastery of the mountain, vitally and nerve-tinglingly alive, on such a really glorious summer's day. The next, with hardly time for realisation, a crumpled lifelessness. There just may have been time for a long second of astonishment, of sickening realisation, then the sunlight was gone for ever.

There are those who, naturally and logically, will say that such a death is a quite unnecessary waste of human life and human endeavour, but, how often does pure logic feature in our lives? There are others who will say that such a death, at the height of one's physical powers and in enjoying the testing of these to the limit, is to be envied. Who shall say who has the right of it? This much I am sure of – those of us who seek the mountains, under whatever pretext, whether hillwalking, birdwatching, shepherding or deerstalking – we are all, consciously or not, accepting some element of risk, which varies according to how foolhardy, painstaking, courageous, apprehensive, or, perhaps more simply, unlucky we are. Should this stop us? Even lengthy experience is no sure buffer against one unlucky slip. It is advised that, also, one should never venture on the hill alone. How many of those who love the out of doors would that advice exclude? For many of us quite illogical human beings there is deep solace and satisfaction in being alone in those solitudes which are still to be found in the Highlands – even though we know that it is only a temporary escapism. The mountains exact only an infinitesimal tribute from the multitudes who enjoy them.

5. *Wildlife in Need*

I have enjoyed and respected the writings of the late Frank Fraser Darling. He it was who wrote 'The animals have outlawed us; they show us fear whether there be need for fear or not' and 'How rarely will a wild animal let us help it, in time of trouble.'

Agreeing totally with those sentiments, it was my ambition to create a sort of wildlife casualty centre and space was available at Torridon. I shall forever be grateful to the National Trust for Scotland for granting me this luxury. Together with the help I was able to give the animals concerned it also enabled my visitors to see these animals and, as a result, to begin to appreciate their existence. For the young visitor there was the possibility of kindling a life-long interest.

Lindy was one of the first wildlife waifs we were able to help at Torridon. She was very definitely the most incorrigibly mischievious and fetchingly frolicsome young animal I have ever experienced; a veritable imp of mischief so that one had to laugh when, instead, discipline was merited. A tiny wild goat kid from Mull, her mother, weak from the birth of Lindy, had drowned in the sea shortly after Lindy's birth. Lindy had been rescued and made the long journey, to Torridon, to be hand reared. She was three pounds in weight, a tiny animated scrap of black and white hair, with already expressive amber eyes. She took to bottle-feeding readily and thrived on it; so tiny was she that in her early days she lived, in a box warmly lined with straw, in the house with us. Perhaps in retrospect, that was a mistake, for she grew absolutely familiar with the house. In the inclement weather of February however it seemed cruel to board such a tiny kid outside and, after all, with no mother to cuddle up to, she had to be kept warm somehow.

Hardly had she become established as part of the household than we had to journey down to Edinburgh for the launch of my first book *Highland Year*. Lindy had to come too, Margaret and I were joint substitute for her mother, the sole fount of her milk diet. Moreover we had already grown fond of her; by joint consent, there was no question of abandoning her to anyone else to look after while we were away.

For the first of our days in Edinburgh we contrived to keep Lindy's presence a secret. Bigger and stronger now, she slept in our vehicle in her warm box; we did not really believe that the hotel management would welcome a juvenile wild goat. We took it in turn to feed her and to keep her company, at intervals. Then someone noticed that Margaret was absent from the breakfast table and I had to admit that she was out buying some fresh milk for Lindy. And who was Lindy? An apposite question, but I'll wager that the inquirer never ever foresaw that the answer would be 'A wild goat kid.'

Lindy was admired and in some cases I believe coveted; thereafter she became a required participant in whatever function was on the agenda. She must be the only wild goat which ever attended a reception held in a plush hotel in Edinburgh. There, I'm afraid, she 'distinguished' herself by wetting all down the front of a red velvet evening gown while being fondled by a motherly lady. She showed her devilish sense of humour by treating me in exactly the same way later, while I sat, wearing my kilt, cradling her on my lap, in a TV studio. I strove hard to remember the stoical qualities of the Spartans of ancient Greece and moved not a muscle, nor uttered audible reproach, with the TV cameras and mikes recording everything.

We stayed a night in Rutherglen with friends before we returned to Torridon. They had a terrier and I borrowed a collar and lead so that I could take Lindy for a walk in Rouken Glen park. For perhaps the first few minutes, Lindy attracted no attention than a child's sharp eyes spotted her 'different' appearance. Thereafter the remainder of Lindy's exercise time was conducted à la Pied Piper of Hamelin. Trailed by an ever-increasing crowd of vociferous children, my ears assailed by shouts of 'Hey mister, whit kind of wee dug is that?' I fled for the car, Lindy extemporising some high-flung, bucking bronco antics en route, to the hilarious delight of the train of children. I believe it was at that point that I began to suspect that Lindy had an eye for the big occasion.

As Lindy grew older, and the weather grew kinder, we gradually got her accustomed more and more to an outdoor life. She was eating grass by now and also whatever else green she could find. I had heard of the omnivorous appetites of the goat tribe; I was to find that this was an understatement, if anything. Out of doors, in an enclosure behind our house, Lindy settled well and almost immediately adopted my tame roe doe as her foster mother. Dainty had no fawn in that year and I'm sure she welcomed Lindy's company; at any rate they were constantly lying together, even appearing to chew their cud in unison. Lindy obviously missed our company though and periodically she paraded, bleating, by the gate of the enclosure, 'asking' to be let out. At intervals she was let out but I really wanted her to get used to being an animal and not a four-legged, black and white coated human. Lindy, however, had a mind of her own and did not always (I

Reflections: Liathach reflecting on Loch Clair.

The Mains; tame deer rising to their feet following a snow shower.

Peat cutting, with Ben Alligin in the background.

The 13-stone Irish Wolfhound which was the subject of a mountain rescue.

These badgers were adopted by me when their parents were killed.

The fox; at home in the rocky ground of Torridon.
An otter enjoying the sun on the beach.

'Marty' the young pine marten spent his early days in the living room at the Mains.

A female pine marten, showing her bright yellow chest markings.

A male ptarmigan showing red eye stripe.

A young tawny owl which had fallen from its nest.

A young heron shows off its long, ungainly-looking legs.

The male dotterel takes his turn incubating; he sat placidly on the nest as I photographed him.

A red-throated diver with her two chicks.

The black-throated diver, incubating but alert.

had nearly written 'ever') agree with what I thought of as desirable. With no way out of the closed gate, she investigated every square inch of the fence which enclosed her, probing, testing, jumping, all in vain. Until the day when, perhaps six weeks after her transfer to outdoors, she appeared at the door of the house, bleating in triumph. Thereafter she could not be kept in the enclosure; she got out at will. I was absolutely baffled until, one early morning, I looked out of the bedroom window to see Lindy nonchalantly picking her way up one of the three-inch wide uprights of the ladder which we had improvised so as to go in and out of the enclosure without having to use the clumsy gate. This ladder, in the shape of a steep inverted V went over the seven-feet high fence. It had an upright at each side, of say three inches wide, while cross rungs of around one and a half inches deep were nailed across these. Up that narrow three-inch wide 'walkway', using the ends of the cleats where they crossed the uprights and, in turn, steeply down the other side, was Lindy's escape route. To Lindy, once this was discovered, this route was elementary. The proverbial sure-footedness of goats notwithstanding, it was wondrous to us to watch as she picked her way, delicately, small hooves clacking, steeply, head down, on the last stage of her Houdini act.

Ambitions, growth, abilities and devilishness all now vieing with each other, her next escapade was to spring from the top of this ladder, to the adjacent felt-covered roof of the garage where, the felt warm on a sunny day, she lay luxuriating, four legs outstretched, yellow-gold eyes half-closed. Her ultimate haven, however, when she felt guilty, remained, unfortunately, the house and this she raced for on the increasingly frequent occasions when she had incurred human wrath. Our door was usually open, or a hard little head barged it open, and she raced from there, hooves clattering, up all three flights of stairs. At the top she made into the nearest bedroom, jumped up on the bed and lay, wide and innocent of eye, on the counterpane. If one pursued her up the stairs (all three flights of them), Lindy had her strategy all worked out. Waiting, innocently esconced on the bed, she let you get clear of the door. A sudden leap from the bed and she was past and, in a diminishing crescendo of hard hooves, she was down and out of the door.

Bob and Jean, next door, had their young grandchildren up from Kingussie that year and Lindy formed a kind of unholy alliance with them. The children went to the village school and every weekday they walked up the length of two fields beyond which, through a strong, high, iron gate, was the road to the nearby school. Up through these two fields, Lindy accompanied them, gambolling and frolicking about, just as they were. At the gate, Jane, the eldest, acted the role of gatekeeper and would order the relucant Lindy to 'go home'. Sedately at first Lindy would go but shortly, since she would never walk where she could run, she would break into a headlong gallop until she was back at the house. When the children came home Lindy was waiting. There was a large, rusty, empty fifty-gallon oil drum behind their house which had been used at one time as a platform from which to hammer in fence posts. Whether by accident or design the children taught Lindy to jump up onto this barrel while they pushed it, revolving, forwards. Lindy (remember I said she liked an audience), the star of the act, graduated from simply keeping her balance while they revolved the drum to actually revolving the drum herself, all four hooves beating a rapid tatoo.

Michael too was accepted as an ally. He was now at the stage when working at the innards of an old Morris 1000 car was both a challenge and a pleasure. While he lay under, or draped over, the bonnet of the car, Lindy sat nearby dozing on an old car seat, taken out of the garage specially for her, or chewing the cud lazily, watching with those supercilious amber eyes Michael working.

In due course Lindy developed sharp, little pointed horns and rather unfortunately, learnt to use these for what in her eyes was 'playful' butting. It was the beginning of the end for, by now, we had a stream of visitors to the Deer Museum every summer. Few of these, understandably, realised that an assault, from the rear, by a rumbustious, sharp-of-horns young goat was 'playful'. The fact that, reprimanded, she then dashed into my house for sanctuary seldom improved matters. The final 'outrage' was when Lindy, truly omnivorous, stole into our neighbour's house and ate all of the luxuriant pot plants which were Jean's pride and joy.

A schoolteacher nearby adored Lindy and had been asking me to let her know if I could ever bear to part with her. The time had come! Regretfully we watched Lindy go to her new home. We heard later that she had signified appreciation of her new quarters by dancing her equivalent of the Highland fling on the roof of her new owner's cherished car. A character indeed and we missed her – her downfall was that her appetite for pot plants was only matched by her appetite for mischief. Something had to go.

A character of similar mould, albeit a feathered one, was a raven which was unable to fly. I had never thought of a wild goat and a raven as being akin yet, believe me, the devilish glint in that raven's bright eye was exactly like that in Lindy's amber eye. He was a handsome bird with glints of blues and greens in his glossy black plumage, and he strutted about inside the enclosure as he wished. His favourite party piece was to sidle up quietly to the feet of any suitably shod female visitor, cocking his handsome head this way and that, wicked eye agleam, and then, suddenly, with one deft move of head and massive beak, untie her shoelaces.

By now I had a couple of badgers, litter mates which had survived when their parents had been killed. One of these became very tame and, with me, utterly dependable. He used to come to the gate of his enclosure on hearing visitors. I would let him out and then he would do his party piece which was simply to take a piece of cake from my hand. Children in particular loved this since I would then pick him up and let them stroke his hard, wiry coat and show them his strong digging paws. The raven, satanic of eye, took to muscling in on this act, scuttling in and off with a piece of the cake which the badger was always punctiliously slow in eating. Came the day when a double act was staged, unrehearsed. The badger, the slow, good-natured, ponderous badger, had had enough. The raven, as was now inevitable, came scuttling in and seized a piece of cake. There was a blur of movement, the raven lay dead, while the badger calmly finished off his cake. I regretted the raven's death but, like Lindy, he had pushed his luck too far.

The two young badgers were from the same litter yet their temperaments were utterly different. One was a rather shy individual who rarely ventured out of the 'sett' which they had dug in their enclosure. His brother (for both were males) was an absolute extrovert, a nice natured animal which I found utterly dependable so that I never ever had any qualms in handling it. We had had all the smaller enclosures completely 'roofed' with strong netting and this was just as well. I had never suspected the badger to have climbing ability until I saw one of mine scale the vertical netting until his nose bumped at the top netting. Badgers are, of course, diggers supreme and in due course a Colditz-type tunnel was dug below the wire. The shyer badger never ventured out; its home and refuge was in its 'sett' in the enclosure; it was well fed and had no ambition to roam. Its brother, bolder in temperament, had an occasional ramble out but always returned to his quarters in early morning; I got accustomed to see him ambling contentedly home in the early hours of the morning.

Because it was so basically rocky and infertile there were no badgers in our immediate area and it seemed quite safe to allow him his occasional ramble. I had no inkling that tragedy was to result! Michael and I, who both regularly handled the badger, had one day to go to Inverness for supplies. While we were both absent the bolder badger came out and, hearing the voices of the children next door, ambled towards them to get, as he thought, his regular piece of cake, from his human friends. Most unfortunately, the first child he came to was young Andrew, who was only about three years old. Andrew, too young to realise that this striped-faced beast was friendly howled his alarm and promptly lashed out with a booted foot. A hard kick smack on his tender snout, when he has expected a welcome, plus a piece of cake, triggered off reflex defensive mechanism and the badger retaliated by biting the offending foot. Howls were redoubled and brought Andrew's mother rushing out and, entirely understandably, she grabbed Andrew and lashed out herself at the badger, who, for the second time, found himself attacked. Again he retaliated, defensively. Jean Robertson, the granny, came rushing out now and she too was bitten, much worse than the others, since the bewildered badger was now thoroughly demoralised and aroused by what he saw as a series of noisy unprovoked attacks. I must emphasise here that this young badger had never had a rebuff or blow from any human until then. The misunderstanding was total. When Bob came in, all were safely in the house. Bob, the most good-natured man in the world who detested to inflict hurt on anything, was prevailed upon by the hysteria naturally prevailing, to go out and shoot the 'ravening monster' and thus died, in tragic misunderstanding, my friendly badger. Had either Michael or I been at home the entire nasty episode would have been avoided, simply by giving the badger a piece of cake.

The Robertsons who might understandably have regarded themselves as the injured party were, instead very understanding and we were thankful to have such good neighbours. I'm afraid that I took a very long time to forget this entire distasteful episode – the badger had been such an engaging animal and by total misunderstanding was now dead. I had been to blame because I had ignored my own guided walks precept, 'play it safe', but it was the badger which had paid the price. After a period of increasingly acrimonious correspondence with Edinburgh I was forced to part with the shy badger also. To be frank, had I not done so, the alternative was to part with NTS.

There were occasions when I was asked to board out animals or birds. An otter, tame since cubhood, proved to be a veritable escapologist. The very first night in her enclosure she found a tiny opening in the corner of the roof netting; this she worked at all night until she had sufficient of an opening to squeeze through. She was soliciting for food at the nearby Youth Hostel at breakfast time and accepted, with alacrity, some fried bacon. By evening she was back in the repaired enclosure and by morning she was away again. She had discovered some staples loose in a piece of spongy wood, dug them out, raised the wire and got out. This time she had made for her real home where they were feverishly constructing a really strong concrete-floored enclosure for her. Since this was not ready she was put in a 'secure' room in a temporarily unoccupied cottage. That night she broke the glass of the only window and was loose once more. Once again she was incarcerated in the cottage, but this time the window was very securely boarded up. She waited for nightfall and got out by climbing up the chimney. And this was the animal I had thought my enclosure would hold securely.

A couple of barely-fledged young tawny owls I boarded until such time as they chose to fly free. The gate of their enclosure was left open every night; there were plenty of mice and voles around and they could supplement their rations with these. Eventually, as I hoped, they flew free. In looking after these I found out that tawny owls had the most excruciatingly painful, needle-sharp talons that I had ever been gripped with. I had had experience with golden eaglé, peregrine falcon, buzzard, kestrel, merlin and sparrowhawk by then. For sheer hair-trigger readiness to strike, needle sharp-penetration and absolute unwillingness to relax their grip, I gave the tawny owls the first prize. I am quite familiar with the advice which tells you that if you ever have the misfortune to be gripped by a raptor that you must not struggle but instead relax completely. This is because a purely reflex action in a raptor is to tighten its grip convulsively if the prey struggles. While acknowledging the admirable logic of this advice I found it rather difficult to achieve relaxed detachment when being pierced by needle-like talons.

Yet another tawny owl was boarded briefly under the alias of a 'buzzard'. I had always marvelled at how little was known, by most country dwellers, of our native birds. I was to have a singular example of this. A friend phoned to ask if I could help him replace a young buzzard he had found at the foot of a tree, fallen from the nest above. I agreed to come across next day to see what I could do. He then volunteered that the nest was near the top of a tall larch tree but that he had a long, extending alloy ladder. He had the young buzzard in the house and this he produced next morning when I rendezvoused with him. The 'buzzard' was a young tawny owlet still in its down of dirty grey. A moment of rather embarrassed incredulity and then my identification was accepted.

It was a very warm day and off we set, in shirt sleeves, the extending ladder with us. The description of 'tall' tree turned out to be a masterly understatement. I am one of these queer individuals who actually relish climbing trees – I did not relish the idea of climbing this one at all. It was very tall, very thin and very awe-inspiring, being one of a closely-planted clump of larches, all of them

.with thin spidery branches. Fine; I had said that I would climb it, I would have to try. In my shirt sleeves, the owlet in a bag on my back, I started up the lengthy ladder which, lengthy as it was, did not reach more than one third of the way up the tree. Reaching its shaky top I forthwith tied it to the tree with the short length of rope which I'd carried for that purpose. Then, slowly I started up that tree my bare arms encircling the rough barked trunk, painfully. Pain or not, it was security I was most conscious of at that time, my security. I inched up, now and then gripping the rough trunk even more tightly when a thin, dry larch branch broke below my foot, or when my bag snagged on another branch. The ground looked a very long way down, Terry's anxious face just a pink blob; I shuddered as I realised the height from which the owlet had fallen – and survived. I really did not want to emulate it. At the nest at long last, up in the swaying top of the larch, there was unexpected disappointment. The nest, (the disused nest of a sparrowhawk, taken over by a tawny owl) was quite empty. We had founded our hopes of putting the owlet back on the idea that it would have nest mates which the owl would have kept on attending. Since the now obviously single owlet had been in Terry's hands for a couple of days we had no way of knowing whether the parents had now renounced their empty nest. My 'epic' ascent had been made for nothing; I just could not risk leaving the tiny owlet to starve. Down I climbed, owlet still in bag, untied the ladder top from the tree, descended its quivering length and put Terry in the picture. He at once volunteered to hand-rear the owlet – he had a couple of young daughters who would cherish and look after it.

The hand-rearing was very successful and later for a couple of weeks I boarded out the owlet while they went on holiday. It was fed on scraps of rabbit, fur purposely left on, and mice and voles and grew into a very handsome red-brown tawny owl. Fully fledged eventually, it roosted, quite at liberty, on a tree by Terry's house, until, one morning, it was gone. I hope I may never have to climb such a tree again, on a mercy mission, even to replace a 'buzzard' owlet.

A pair of wildcat kittens, fist-sized, but fiercely defensive (not offensive) arrived. Their mother dead, I was asked to try rearing these. Defiant as the tiny kittens were, they settled down well. Eventually I was able to handle them with impunity yet I must confess that I never ever felt that I had gained their confidence and to that extent I was not successful. The kittens grew to maturity and appeared to thrive. One dark night some ill-intentioned, ill-motivated, individual opened their enclosure gate so that they could 'escape'. To little purpose; they did not want to escape; they were, instead, bewildered. Neither of them, left the area which they had come to regard as home, where they had abundant food and shelter. I found them crouched in a corner of the main enclosure and was able to lift and return each one without scratch or scathe. Alas, first one then the other, months later, died of what the vet said was a kidney infection which was often inherited by felines. I was sorry but, in a way, perhaps it was for the best; they had tolerated rather than trusted me, always remaining aloof. Perhaps it is a facet of feline nature. Yet, again, on being set free on that above-noted occasion, they had not departed for the wide open spaces, as they could so easily have done.

Foxes were quite unpredictable; they varied in temperament enormously from being incurably furtive and suspicious to the other extreme of being overwhelmingly demonstrative. However they were never aggressive. In many ways they were extremely dog-like in behaviour, in burying uneaten scraps of food, in expressing feelings by whining and wriggling their bodies, and by wagging bushy tails when happy. Their life span also approximates to that of a dog; I had a dog fox for thirteen years and a vixen who died at fourteen years. As with dogs, old age brings worn and hence inefficient teeth; meat-eaters require efficient teeth to eat adequately. Fox and dog, however, will not interbreed. The fox is never aggressive (one must, of course, except a rabid fox) it prefers evasion to aggression. I could always remove a piece of meat from a fox while it was actually eating this meat; I would not back myself to be able to do the like with a domestic dog without getting bitten. It is probably the similarities to the dog which has encouraged so many people to try to tame and hand-rear an orphaned fox cub. A laudable enough act but the scales are heavily weighted against it being successful. The majority of 'rescued' young foxes do not ever trust the human who is attempting to tame them and they are always looking to escape. For this reason they can be frustrating to the point of heartbreak; one always has to remember that the fox has virtually no friend in the Scottish countryside.

The most dramatic and unexpected arrival ever to come to the Mains, seeking help, was a golden eagle. Dramatic because it came literally out of the blue, entirely of its own volition; unexpected because one just does not expect a free-flying apparently wild golden eagle to suddenly land at one's house. Since I was known to have been studying the eagle for many years there subsequently arose murmurs about black magic and occult arts. Seriously however no one was more amazed than I was to have this happen.

Our nephew Douglas was staying with us and he had gone for an evening stroll along the loch shore. At 10.15pm he arrived home, in some excitement, to tell us he had just seen a huge bird fly in, low, from the loch and land on a fence post. Almost without thinking, I said 'Probably a heron!' Douglas looked his scorn. 'I know what a heron looks like' he said. Sceptical but curious I went down towards the loch with him. There, perched on a post, and looking enormous in the fading light was, quite unmistakably, a golden eagle. A moment of utter incredulity and then it took wing but, instead of flying away from us, it flew up towards us, passing us about head-height. At this height its sheer size and length of wingspan was truly awe-inspiring, one could well imagine the terror of any prey species as this enormous winged spectre of doom stooped at it. It landed in a bold flourish of wings on a post of the deer fence by my house. Two of my hinds spotted it land and a curious yearling minced forward with hesitant steps for a closer look. The other hind, her mother, had never seen an eagle in her life but she showed no hesitation; rushing by her yearling like a fury she rose to her hind legs and lashed out with flailing forelegs at the eagle. Routed, the eagle took wing again and this time landed on the chimney of the house next door which was now unoccupied. At that time I had been watching and recording, at the eyries of the golden eagle, for more than twenty years and if anyone had ever said to me that one day I would see a golden eagle perch on a chimney pot I would have doubted his sanity. Lest anyone should doubt mine I had Douglas and Margaret as witnesses. More than that, I dashed in for my camera, got the telephoto lens on and, back out, essayed a photograph of this unique sight. The light was now so bad that I needed a two-seconds exposure time and this I tried to compensate for by holding the camera steady against the corner of the house. The resulting photo is by no means a pictorial work of art but I'll guarantee that it is the only one in existence of an eagle on a chimney pot.

You will understand that it was difficult to get to sleep that night and next morning, despite being quite sure that the eagle would have gone I got out of bed at first light. Typically, it was a miserable, grey, weeping morning of that type of soft, fine, west coast rain which had every blade of grass bending under its weight of moisture. And there, perched on a fence post, screened by a short length of hedge, was the eagle, still with us, and looking just as moisture-laden on its feathers as was the grass. This was not a wild eagle – that was the only conclusion. Yet there was no sign of leather jesses on the legs, nor did I know of any falconer in the area. Still speculating, I went in for breakfast. While I was breakfasting this incredible eagle flew nearer to the house, to a fence post behind the house. There, on its perch, it watched a couple of rabbits, playing around, quite close to it, and made no effort whatever to catch one. Live rabbits, running about within easy range, obviously meant nothing to it. I re-appraised the situation and, going out, re-appraised the eagle also. It had no fear of me or of close proximity to buildings; it was obvious, abundantly obvious, that it was not going to fly away. Its crop, at the junction of lower throat and upper breast, was absolutely flat, no sign at all of recent feeding. The penny belatedly dropped – the eagle was hungry, possibly starving, and it seemed incapable of catching prey for itself. It had come to my house impelled by the strongest of motives, food, so that it could survive. As I watched from my back door two common gulls arrived and began, calling raucously, to dive-bomb the perched eagle. They almost brushed the eagle's head each time they stooped and the eagle cringed abjectly and drew its head in between its 'shoulders' at each stoop. I had seen gull before now, as prey at an eagle eyrie, but this eagle exhibited no sign of attack or even of retaliation.

I went to where I knew I could get a rabbit and called at the school to ask if any budding bird watchers would like a close-up of an eagle. Then, with dead rabbit, I went into the field beyond the fence, followed by the excited children. Close to the still-perched eagle I threw the dead rabbit onto the grass. Just as soon as it had landed so did the eagle push off from its perch and with masterly precision, pounce on it. Legs astraddle, one huge, steel-taloned yellow foot at the neck of the dead rabbit, the other gripping the rump, it reared up to full height and, head back, its amber-brown eyes glared defiance at anyone and everyone, to come and try to take its prey from it. Seconds later, head and hooked beak bent to tear its first beakful of fur from the rabbit, it began to feed ravenously. The eagle had indeed been starving yet live rabbits, scampering about, near to it, had triggered off no food impulse; dead rabbit, immobile, was however a familiar object.

Half an hour it took for its meal, in that time the entire rabbit, except for the discarded entrails and littered patches of fur, had vanished. Meal ended, the crop of the eagle was no longer flat, instead it bulged out in a pronounced swelling curve, rather like the proud figurehead of some old time galleon. The gorged bird then sat in the field for five and a half hours, satiated with rabbit flesh; then for yet a further five hours it meditated, perched on another fence post, totally immobile. I had long known that the eagle on the hill spends a lot of time just perched, immobile, digesting the food its crop holds but this lengthy period, a total of ten and a half hours was a revelation. Little wonder it was rare to see an eagle on the hill unless it was on the wing. An eagle, perched on some sheltered ledge, immobile, digesting food, would be almost impossibly difficult to see.

Next morning, as expected now, 'my' eagle was still present, perched on a post some 150 yards from the house. Getting another dead rabbit I waved it in the air so as to attract the eagle's attention. It was a mistake! The eagle pushed off, flew straight, low and fast at me and, swinging fast in behind me, probably struck out at the rabbit, missed, and sunk a taloned foot into what I shall politely call my right haunch. In an immediate pained and aggrieved reflex action I swung the rabbit, hard, in a curve behind me and knocked the eagle off, at the same time continuing the swing to let the rabbit fly from my hand to the ground. Losing grip and interest in me the eagle at once transferred it all to the rabbit. Plucked out fur flew from its hooked beak as it curtsied and came upright, in jerky, flesh-tearing-out movements. A timed half hour later the rabbit was gone; again the crop was jutting out, prow-like, and again the eagle was immobile for, this time, a total of eight and a half hours. Towards the end of this digesting period, in a sudden, violent shaking motion, the eagle ruffled out every feather so that, momentarily, it looked like a broody hen, an almost circular shape of dishevelled, loose feathers. Within seconds the dishevelled look settled down to sleekness once more while a few wisps of the thistledown-like inner white feathers floated gently away on the light breeze. I had often found such thistledown wisps of airy-light gossamer caught on heather and rock. Here was the explanation. A brief preening session ensued; it was revealing to watch the delicacy and precision with which the huge break was employed to gently caress into place the long primary feathers of a half extended wing.

On the third day there was a variation; after two good meals the eagle was now perched on a roadside telegraph pole. A buzzard began mobbing it, diving down, mewing, almost touching the motionless eagle. So incessant and so vigorous was this attack that the pestered eagle at last flew and landed on a flat across the road where it perched until late afternoon, in full view of the busy road. The situation had now become complicated. Where it had landed near the main road the eagle was beyond my personal protection. At best it was at the mercy of any and every sightseer; at worst it was at the complete lack of mercy of any ill-doer. Nor was the danger all to the eagle's well-being; it had already proved conclusively that it was no respector of humans, in its puncturing of my rump. I however, at the expense of some pain and outrage, was well able to deal with that sort of thing. What of some unsuspecting youngster however, gripped by these potentially lethal talons in a less well-padded place than that of my lacerated rear? I pondered on three choices of action: (1) Do nothing and trust that it, when hungry, would fly back to the Mains; risky, too many things could go wrong; (2) Take a rabbit to it and, after it had demolished this, it might decide to fly into the hills but this I felt was most unlikely. The eagle had demonstrated that it could not catch prey, that it needed human help. (3) The third course seemed inevitable; to go over, catch the eagle somehow and take it back over to the Mains, where I would keep it, safe, in a roomy, roofed enclosure until a solution turned up. This, finally, I decided on, the eagle would be safe then, and it would be guaranteed food; I would hear soon, I felt, that it had a human 'owner' and he would hear of the eagle which had chosen the Mains to land at in search of food, and of understanding. Although I was reluctant to imprison the free-flying eagle, I made haste to take action.

Douglas was still with us, and I had a long handled net which I had used at one time to catch deer calves for research purposes. I drove across with Douglas to the perched eagle. It inspected us with a cold amber eye and moved not one bit. I asked Douglas to try and hold the attention of the eagle while I worked around behind it and tried to get the net over its head. It worked, but it also triggered off an awesome display of unbridled rage from the bird which, a moment before, had been so placid. The eagle was now transformed into a wildly indignant fighting fury, lying on its back and striking out viciously through the wide meshes of the net with both beak and furiously clutching talons, eyes glaring wildly in magnificent but impotent wrath. Wrapping an old coat around the netted eagle left it in darkness and quietened it somewhat, but we lifted the animated bundle with due caution lest a striking talon got loose, and transferred it across to the van. Back at the house we carried it to the netted enclosure where freeing the eagle from the net, in which it had contrived to get itself thoroughly entangled, proved a major problem. I could never have managed to free it single-handed; even with the aid of Douglas it was a very difficult task. Both of the eagle's wing 'elbows' were through separate meshes, as also were both the huge and potentially dangerous feet. The head was, somehow, through two of the meshes. I freed, with difficulty, both wings first, keeping the head and beak 'blindfolded' as I did so. Next I freed one taloned foot while Douglas held the shrouded head. Then, freed foot firmly held in one hand, I contrived to free the other foot. Both feet tightly held in one hand now (I dared not allow the inexperienced though willing Douglas to hold these dangerous 'weapons') I somehow managed to manipulate head and beak free of one encircling mesh and then the other. Warning Douglas to stand well clear I now pitched the freed eagle gently forward onto the grass. Amazingly, the unbridled fury of only minutes ago seemed forgotten; docile now, if indignant still, the eagle jumped up onto the rock and shook out all of its outraged feathers into brief dishevellment, before reverting to its normally sleek outline. I went to fetch it a rabbit, by way of propitiation, and this it accepted magnanimously, thereafter lapsing into the usual meditative semi-trance. We, thankfully, left it thus.

For the next few days I varied the eagle's diet by sundry items from the deep freeze. Then at my door there arrived Dick Balharry who had been at Beinn Eighe Nature reserve when I came to Torridon first. He had later been transferred to Aberdeenshire. There he had had 'my' eagle handed in to him by the police. They had found it weak and starving, the rotting remnants of leather jesses on its legs, and, rescuing it but unable to do anything else for it, had taken it to Dick. It was, he said, not far from starving to death, only about half the weight which it should have been. He looked after it and, very gradually, it had regained fitness. This had taken months to achieve, no wonder the eagle had no fear of humans. The next task was to prepare it for life in the wild, in the mountains where eagles are at home. This had been much more difficult; however in due course Dick was sure that the eagle would be able to fend for itself when flying free. Since he knew Wester Ross to be good eagle country he chose to take it right across Scotland from Aberdeenshire to release it in Wester Ross. He had done this some ten days before the eagle arrived at my house. The rest you know; the eagle, in its second experience of near starvation, had, perceptively, decided that I could feed it. Dick took it away. I believe it was eventually decided that its future lay in a wildlife park; it was altogether too well attuned to human ownership to be able to cope with a wild existence. The 'black magic' interlude was over.

6. More Wildlife Waifs

A gale-force wind was howling its banshee way down Glen Torridon with frightening speed, savagery and seemingly inexhaustible fury, as if indeed it was pursued and possessed by a thousand shrieking demons. With it it carried a multitude of horizontal spear points of rain, salt spray from Loch Torridon and, at intervals, large and extremely hard hailstones. These latter, driven horizontally, stung my exposed face almost beyond endurance, so that I had to turn my back on them. Yet it was a chillingly cold day, the wind, off the Atlantic, driving right through one.

In the narrow glen the waterfalls, while attempting to plunge downwards on their rock faces, were unanimously defying gravity. In most cases they were being swept out in a ragged horizontal tail, or even, in others, whisked vertically upwards in a cloud of white spray, according to the whim of the raging, mountain-constricted, Atlantic-spawned gale. Every blade of grass and every tuft of heather was bent in supplication before the screaming wind and bowed under the weight of chilling moisture. The few trees near the house were cracking and groaning in torment, their leafless branches bending and whipping before the savagery of the gale. The long inner arm of the sea loch was a forbidding steel-grey expanse, broken up by an endless succession of wind-tortured, white-tipped waves racing each other, in long lines, to the rock and shingle of the shore. The very house shook and shuddered in all three storeys, strong and thick-walled as it was; its large windows rattled incessantly in their frames under the devil's tattoo of sleet and heavy rain, a furious onslaught which one felt must surely eventually breach them. It was awesome, it was brutal; it was Torridon!

On such a day in winter one realises a little, just a little, of what the red deer, on the hill every day of the year, have to endure simply to survive. They are denied the earthy hole, or rock cranny, of fox, badger or wildcat; denied even the woodland shelter of roe deer; denied the human shelter of roof and fireside. It is only their supreme instinct for achieving the maximum shelter (in hill ground where shelter is minimal), from the driving wind, their hardy constitution and their sheer unquestioning capacity for endurance, under the most adverse weather conditions in Britain, which enables red deer to survive each Highland winter. And not only to survive, but to thrive, in areas like Torridon, which, in this savage, unrelenting uncompromising mood, can rank with the bleakest, most inhospitable, hostile country in all of Britain.

It was on just such a day that a roe deer doe of mine, and a crippled roe buck, got out of their enclosure, the gate having been blown open by the gale. This was at 2.30 pm and I had had a morning session (indoors I need hardly say) with a BBC TV unit who were now having lunch. Roe deer detest gale-force winds and these two were completely demoralised by the weather. I managed to lure the lame buck back into shelter with a bucket of feed, the doe was following nicely when a howling gust of wind wrenched a large branch from an ash tree, with a rending crash. The frightened doe took to her heels over the fields, seeming to float spring-heeled in the gusty wind. At first I thought I might coax her in on my own, with the bucket of feed as bait. How wrong I was! Time after time I had her near the gate – a tremendous gust of wind would come, a slate would rattle down from off the roof – another ash tree would creak, alarmingly, in torment, and the doe was away again. I became hoarse, cursing uselessly but satisfyingly, that insentient wind. I was already tired out when Michael joined in the pursuit but even then, our forces doubled, the weather and the now highly alarmed state of the doe, meant that we were thwarted time and time again. Two of the TV crew joined us but the fleet-footed roe doe beat us. Before very long, George and Simon, the two TV chaps, were just as leaden-legged as Michael and I were by then.

The seemingly tireless doe got through the last fence and then she was on the shore, at the head of the sea loch. I made a final dash to try to head her off but she eluded me easily and made for the sea. Michael and I followed as fast as our legs would allow us, but wind, rain and spray got in our eyes and we lost sight of her. Surely she would not risk that welter of wind and tide-impelled waves, which were piling in, ceaselessly, to the shore? We'd just decided to split up and each search different stretches of the shore when an excited shout from George halted us. He was waving his arms excitedly and pointing down the loch. We looked and my heart sank for there, well out among the waves, was the doe, only her small head visible as she tried, weakly, to return to the shore. I began to get rid of my jacket and then my thick jersey and caught a glimpse of Michael doing likewise. We both went wading in, about one hundred yards apart, so as to have the doe between us when we reached her. The seawater was bone-chilling in its temperature. It was ankle deep, calf-deep, knee-deep and then, suddenly and alarmingly, it was chest high, as we both, slowly and laboriously, got closer and closer to the doe, keeping our distance a hundred yards apart. Out beyond the struggling doe we turned inwards so as to follow the doe as she was swept shorewards. And then, in a nightmarish sequence of slow motion, I saw Michael appear to go out of his depth. Suddenly only his face was visible, mouth open, gasping for air and his hands flailing as he essayed a clumsy sort of 'dog-paddle', for he could not swim. It took me a long, long time afterwards to forget that desperate expression on his wave-battered face. Doe completely forgotten I forced one leaden foot after another in a desperate race towards Michael, dreading, with each step that his face would disappear in that welter of grey, gale-swept seawater, before I could reach him. Almost in a fear-filled haze myself I suddenly became aware that Michael had regained his feet and, almost within arm's reach, was mouthing at me that he was 'all right'. He took a slow step or two towards me and relief, really indescribable relief, swept over me. Words, in that wild wind and wave-torn scene, were not needed; we both began to make shorewards after the doe.

Nearer to the doe than Michael was, I had almost reached her, now in waist-deep water, when an extra large wave knocked me over. In the combination of hurrying waves and gale-force wind, even though I could swim, I found it well nigh impossible to regain my feet. I knew then, with frightening clarity, just what had happened to Michael. He had not gone out of his depth; he had been, literally, like me, blown off his feet. In my consequent panicky floundering, at times almost crawling on the bottom, I swallowed quite a lot of seawater but managed to keep myself pointing shorewards. George, of the TV crew was beginning to wade towards us when, in the shoaling water, I contrived to get to my feet. More by luck than good judgement I managed to grab the absolutely water-sodden doe as she turned away from George. I found then that she was so weak and so heavy with seawater herself that she made no attempt to struggle. Michael splashed out just behind us and immediately said he would go up to the house and fetch the Land-Rover down. Stumbling, slipping, staggering I carried the dripping and virtually inert doe up the shore. Without George, who was now quite literally half-supporting my sodden form as I wavered on, I could never have made it. Simon now arrived with the outer clothing which Michael and I had divested ourselves of when we went for our 'dip'. The Land-Rover, thankfully, arrived, and we got the doe in the back and unashamedly I lay down on the floor beside her. I managed to stagger, dripping, into the house while Michael carried the doe in. Moreover, I managed to find the bottle of whisky I had in the sideboard from which Margaret poured out mammoth drams for George, Simon, Michael and myself. Four of us, squatting on the rug in front of a blazing fire, together with a sodden roe deer, is hardly an everyday domestic occurrence. Margaret was wonderful, in her acceptance of it all, trying to help me to squeeze some of the water out of the doe's coat, which, of course, in logical sequence saturated the rug. She made tea for us, hot, sweet and tasting like nectar, whilst also fetching an odd selection of dry clothing for us. Exactly what happened next I have never been sure! One moment I was chatting away, voluble with relief, whisky and hot tea, the next I was dimly aware that I was being carried off to bed, clothes dragged off, pulse taken and hot water bottle administered. I could hear things going on around me but I could not respond myself. I was deathly cold and shivering uncontrollably. The doctor duly arrived and diagnosed utter exhaustion and some hypothermia. I slept marvellously, possibly helped by the doctor's sleeping tablets and was right as rain by morning, as were my accomplices, Michael, Simon and George.

The errant doe, her attempt to reach America unsuccessful, spent three days and nights lying in our bathroom, the warmest wee room in the house, with the immersion heater and the wall heater on constantly to keep her warm. We, mere humans, plodded up to the top floor to use the spare and rather spartan bathroom which was there.

What a scoop for TV that wind and waves rescue would have been, and we had had a crew on the spot. Unfortunately the atrocious weather made it quite impossible to film even if any of us had ever thought of this at the time. Life in Torridon was not always of tranquillity supreme as so many of our summer visitors assumed.

The next casualty to feature at the Mains was, thankfully, not nearly so demanding. My eldest son, Lea, had been asked to come to the aid of a pine marten which had been illegally caught in an outlawed gin trap. The pine marten had sustained a broken leg and he asked if he could bring it over to me. I immediately agreed and, some hours later, Lea arrived, complete with the pine marten in its cage. Meanwhile, pending its arrival, I had hastily improvised temporary quarters for it in an old tea chest. The caged marten was neither vicious nor aggressive, it made no attempt to bite as I moved my hand near to the wire of the cage. Very slowly and gently I thrust my hand in from the rear and slid it over the wee creature's back until I could grasp it just behind the head. In fact I treated it rather like a ferret, and luckily I had always been fond of ferrets and was well used to handling these sharp-toothed beasties. I quite anticipated a defensive bite, in natural apprehension, from the pine marten, but no, this it made absolutely no attempt to do. Extracting it gently, taking care not to inflict any more pain to the injured leg, I found that it was a male, amazingly quiet and docile despite the trap-inflicted broken leg. On request, Lea held it for a moment while I took its portrait. On examination the leg was indeed broken, just above the mid-joint. The wound however had now stopped bleeding and, since it was late in the day, I decided to leave the pine marten to have a rest overnight while I phoned the nearest vet in Dingwall. The first vet I tried was extremely sympathetic but advised me that the best thing to do was to put the wee animal painlessly 'to sleep'. I thanked him and tried another vet. This time I struck gold: 'Bring it in first thing in the morning and we'll see what we can do,' the voice said.

Early next morning I was off on the fifty-mile journey to Dingwall, the pine marten comfortable in a smaller wooden box, on the seat beside me. Mr Macintyre the vet ushered me in right away. He also asked if I would participate in the act by holding the injured pine marten since neither he nor his assistant had, up to then, had any experience of treating a wild pine marten. I readily agreed. On examination he said that the injured leg would have to be amputated. He doubted if splinting it and putting it in plaster would work since the pine marten would inevitably try to bite this off. He was quite astonished, and probably relieved, at the docility of the 'wild' pine marten. While I held the pine marten he administered an anaesthetic, 'Enough to put a comparable-sized animal, a cat, say, to sleep' he vouchsafed. While the anaesthetic was taking effect he busied himself by assembling his 'tools', his assistant helping, and in sterilising these. I was immensely impressed by the care and thoughtfulness taken throughout the entire treatment of what was 'only' a wild pine marten. The marten now absolutely dead to the world, the maimed leg was cleanly cut off; the sharp end of splintered bone was then trimmed off and shortened to enable a good pad of flesh and skin to be pulled over the stump and stitched into place. Before this took place and skin to be pulled over the stump and stitched into place. Before this took place however I was intrigued to see a kind of miniature hairdresser's clippers employed to trim the hair of the coat away from the wound. The time, skill and patience required to put in the necessary number of tiny stitches needed to secure the perfect protective pad over the stump of the leg was, once again, a revelation to me and I was left with profound admiration for the dedication involved. Work complete, I paid the embarrassingly small professional fee asked for, said my sincere thanks, expressed my admiration and we were away home to Torridon. I was nearly there when I heard a scratching and rustling from the box and the pine marten's head, bleary of eye as any reveller after the night before, poked, shakily, over the edge of the box, before subsiding for another nap. Once at home, I immediately set a mousetrap (we had always an ever-present population of woodmice at the Mains) and shortly had a nice fat mouse. This the now speedily-recovering pine marten relished; he may have lost part of a leg but he had regained his appetite quickly. He also had some fruit cake but most of all he showed great liking for some 'biscuits' which Margaret, rescourceful as ever, had baked from scraps of fat and oatmeal.

I kept him indoors for ten days, in a big roomy wooden chest which had been a corn bin, with ½" wooden sides and lid, and this I kept in the concrete-floored deer larder. He demonstrated full recovery by biting his way out of what I had thought was an impregnable wooden chest, ½" walls or not. Fortuitously, he achieved better exercise by roaming the full extent of the roomy deer larder, and he thenceforth retired to the box only to sleep. The main food available was rabbit and this he seemed to relish. Recovery complete, in due course he was transferred to one of the outdoor pens where his observed agility showed me that his visit to the vet had not left him with any tremendous handicap. In course of time he returned to the wild. I was confident that he could cope and quite sure that, if unable to do so, he would return to the house. It may well have been this pine marten which was the culprit who, later, regularly visited tents, under cover of darkness, in the nearby camp site and who demonstrated a wide ranging appetite in the titbits filched thereby. My admiration for vets, and my admiration for the strength of the teeth of the pine marten, were alike increased by this rewarding interlude.

'Lucky' was an exceedingly appealing roe fawn; she came to Torridon as a result of a 'phone call S.O.S. from a good friend in Morayshire. My caller really brought home to me the problem posed, yet again, by modern techniques in our mechanised agricultural practices, this time of horrific injury to roe deer resulting from the use of high speed machinery used in cutting hay and silage.

Morayshire, on the east side of Scotland, could hardly be more different from Torridon. There, woodland plantations alternated with the broad acres of top quality farmland, and, not surprisingly, since those are ideal conditions for roe deer, Morayshire has also a high population of top quality roe. The woods give them shelter and the farmland provides feed. Roe deer are exceedingly fastidious and selective feeders and while we see them feeding out on the fields they are most often feeding on the 'weed' species which grow amid the grass. What are weeds to the farmer are often attractive food species to the roe deer. Unfortunately, in late May/early June, many recently-born roe fawns, tiny, dappled young things (they only weigh about 4 lb at birth), will be left lying, snug in the cover of the long grass, by the doe. The doe cannot visualise that this cover is shortly to be slashed down by the high speed multiple knives of silage or hay cutters. Coupled with this, the over-riding instinct of young roe is to lie tight when danger threatens, consequently, on the roar and clatter of machinery approaching, the roe do just that – lie immobile and virtually undetectable. The centuries-old survival technique of lying tight does not take account of modern high speed machinery and, in a trice, an entire roe family group may be, literally, minced up, without the machine operator being aware of this, for the mother will also lie tight to 'protect' her young.

My friend, Jim, had 'phoned me to say that just such a tragedy had occurred near to his home, and, by some whim of fate, a female fawn, from a party of mother and two fawns, had miraculously escaped injury. His knowledge of roe deer being known in the area he was asked to try and rear the surviving fawn. This, believe you me, is never an easy task; the milk of both red deer and roe deer is much richer in content than the milk of a dairy cow, yet most amateur 'hand rearers' mistakenly dilute cow's milk and make it even weaker. However, the immediate appeal of the tiny dappled sprite was impossible to resist and Jim decided to try it. A corner of his not very large garden, in the town of Keith, was segregrated for the fawn's nursery, and Jim, for very obvious reasons, called her 'Lucky'. Somehow, somewhere, Jim contrived to get hold of a baby's feeding bottle and amazingly the fawn took to this alien teat, and alien milk, quite quickly. At around three weeks old she began to experiment with the fascinating greenstuff which surrounded her and that is when Jim's luck ran out. The greenery in her restricted bit of the garden soon dwindled; Jim's wife was fond of her flowers, and of Lucky too by this time. Regrettably Lucky proved to be equally fond of some flowers but for a very different reason. Matters, horticulturally, went from bad to worse, to crisis, and Jim found that the streets to Keith were not paved with flowers. His gardening cronies were beginning to shun him; it was then that, in late July, he telephoned me.

A day or two later and Lucky arrived, with foster parents, by car, lying in the back, surrounded by greenery, Keith's last contribution to her taste for this. Both Jim and his wife were very obviously sad to part with her, such is the abundant charm of young roe deer, but they had also performed nobly in getting the fawn to this stage – now it was over to me. My problems were not long in starting. Liberated into a roomy grass-grown enclosure, Lucky refused, absolutely refused, to come near to me. This 'new' human was utterly unacceptable; not even milk-hunger would bring her to me. Since it was obviously unthinkable to chase her around and clinch with her every time I wished to feed her, I had to think of a way round the problem. Apprehensive, uncertain, I tried putting a bowl of milk in her pen and leaving her to see if she would drink this. I watched from a distance as she approached the bowl, circled around it, sniffed it, then began to sample it greedily. I gave a silent cheer, alas, too soon, for, in her now aroused milk hunger Lucky was too eager and, almost immediately, tipped over the bowl. Back for more milk, and a garden trowel. A suitable recess was dug for the bowl and I left once more – no milk for my tea that evening. I was obviously not going to be allowed near enough to handle Lucky; never mind, the more independent she kept with me the better it would prove when she chose freedom eventually. Next step was to establish the species of greenstuffs she favoured since there was only grass in her enclosure. Trial and error established that, in leafage, she had very definite preferences. *Salix*, that is the willow species, proved a firm favourite. Since roe deer are said to be fond of bramble (blackberry) leaves, I tried that also. This was accepted but it was only the tender tips of the spiky tendrils which were eaten. Leaves of ash, gean (wild cherry) and rowan were rather more gladly accepted, as were leaves of apple and plum tree. Leafage of birch, alder and hazel were only eaten grudgingly and were very obviously second choice. Unfortunately it was an accepted fact that roe deer relish cultivated roses (Jim's wife sadly confirmed this). I had first-hand verification of this when I brought in some rose foliage for Lucky. Obviously, very obviously, this was the first choice. Not only leaves were eaten, but then the flowers went down also, nothing was left but bared, thorny stem. There is of course, a much smaller limit on the number of 'surplus' cultivated roses available in a sterile-of-soil West Highland village, than in a rich-of-soil Morayshire town. Roses became available only as a much-relished treat. Wild roses, briar or dog roses, were, I found, to my astonishment, utterly unacceptable. Nothing suited, in the way of roses, but the very best.

I gradually weaned her by first of all trying her with a kind of muesli mixture which I myself favoured for breakfast. I thought it rather unlikely that she would like this; unfortunately I was proved wrong and the muesli container was quickly emptied. The local shop had none in and I had to change my breakfast routine; Lucky had coarse oatmeal and enjoyed that also. All in all, on her mixed and at time experimental fare, Lucky waxed fat and flourished whilst I lost some weight. Later that autumn Lucky was given the freedom of the large enclosure which she shared with my red deer. Still later my reds, and Lucky, who by now thought that she was a red deer, were given the freedom of the fields. Until the following June Lucky remained with my red deer, changing coat from her first dappled one to the immaculate, thick dark coat of her first winter and, finally, in the ensuing June, to the elegant, trim, rusty-red coat of the summer roe deer. She never did become hand-tame and she would never allow her head and ears to be scratched as my red deer would. Nevertheless she left an emptiness behind when, her first mating season on her, she slipped through the outer fence one day, and thereafter, found her own leafage and wild flowers, and, probably, her own buck.

A most memorable and endearing wee creature, a young pine marten, Marty, arrived, heralded by the usual 'phone call, in early July one year. I believe it got separated from the family den, somehow or other, and could not find its way back. It was found wandering across a tarmacadamed road, helpless, starving, quite near to death in fact.

To begin at the 'phone call, however; this came in late afternoon, from Mrs Calcott, who lived about a mile from Shieldaig, on the coastal road which continued around to Applecross. Her husband had been driving along that road when, amazed, he spotted a pine marten wandering aimlessly on the road ahead. In fact so unco-ordinated were its movements that his first thought was that it was blind. His teenage son was with him and he stopped, let his son out, and he caught the pine marten easily, for it made absolutely no effort to evade capture. They took it home, put it in a box by the fire and tried it with some milk and some cooked sausage. Perhaps not surprisingly, the pine marten rejected both. It was at this point that they rang me. Over I went, intrigued by the story, and found that the pine marten, obviously a young one, probably about three months old, was dreadfully thin and weak. I told the Calcott family what the situation was and was asked then if I could look after the pine marten. I thanked them for the trouble that they had already taken and said that I would help. And so Marty arrived, driven home, without fuss or protest, lying curled up on my lap wrapped in my red woolly balaclava to keep him warm. Permission being granted by the chatelaine I took Marty into the house and introduced Margaret to yet another enforced lodger in our house. I managed to get him to eat some bits of rabbit flesh and then he curled up and went to sleep on a bean-bag, as if such a sleeping place was what he had used for all of his three months of life so far. Marty proved a most engaging little creature who gave us implicit trust from the very start and yet also some dreadfully anxious moments 'ere he was back to complete fitness.

On 16 July I took Marty to Roy Peacock, a friend of mine, an exvet, who had now retired to live the life of a crofter at Upper Diabaig. Roy as ever, was very helpful and he advised an antibiotic injection and also a vitamin one. He also gave me some glucose solution to be given orally. Next day, frankly, I regretted the injections which we had agreed on. Marty was wandering around our big living room as if drunk, his hindquarters, in particular were quite unco-ordinated. He kept blundering into the furniture, falling, getting up again and resuming his restless prowling for he seemed quite unable to rest or relax. I had to force-feed him with small scraps of rabbit liver, and with the glucose solution, fed into his mouth via a plastic disposable syringe. Margaret suggested an egg, beaten up with sugar; this I tried, feeding him periodically throughout the day with the aid of the syringe. It was midday next day before he showed signs of returning to normal – we both breathed huge sighs of relief and I began to think of an outdoor run for him, provided he maintained his recovery.

Some days later, by then having constructed a warm, weather-proof box, I introduced Marty to the large outdoors enclosure where he was now to spend the daytime hours. For at least a month yet I was going to have to have him back into the house to sleep – the nights were chilly for such a wee creature on his own. He had the time of his young life exploring these new quarters but, rather to my surprise, he showed that he did not like warm sunlight. He would not, for instance, have a nap lying curled up in full sun, but, panting, open-jawed, and exhibiting some distress, he sought a shaded spot to sleep in.

Indoors every night, Marty had a great time, as if glad to be inside once more. His activity was incredible for a brief mad spell, then, tired out, he had a sleep for an hour or so. The frenzied bout of exercise included racing around the room, jumping up on chairs or couch, bounding over them, and making mock, ferocious attacks on a jacket of mine which hung over a chair back. Another favourite for these mock attacks was a draught-stopper, in the shape of a long, thin, hairy counterfeit dachshund, which lay by the living room door. Marty had tremendous fun with this, pouncing on it, 'killing' it (a shake at the back of the neck accompanied by the pine marten 'growls') then, dragging the 'dead' dachshund all round the room and finally, rolling over onto his back, clutching the dachshund to his chest and going to sleep thus. Since our living room had a good thick carpet on its floor, Marty was liable to go to sleep anywhere on it. After a bout of furious action he would, quite suddenly, 'switch off' and be soundly asleep in seconds. These were obviously the sort of growing-up exercises which would have taken place with his litter-mates in the den from which he had gone astray. They were beneficial to the well-being of Marty and great entertainment for us, together with some interesting insight into marten habits.

By now Marty was drinking his daily beaten-up egg from a dish, relishing this method, for he had never enjoyed the necessary adminstration by syringe. Nor had I enjoyed this, so that now we were both happy. This diet was now enlarged to include raw mince, cooked chicken scraps, rabbit and also ox liver from our butcher. Marty would take small pieces of his meat from my fingers, amazingly gentle, never over-hasty or greedy, in doing this. Intriguingly he grasped each piece proferred to him in his cheek teeth, (the equivalent of our back teeth) rather than in his front-of-the-mouth incisor teeth. There was not a vestige of viciousness in his behaviour even when grabbed at night, to be put into his sleeping box. Altogether he was a most appealing and rewarding wee creature, and an education in the behaviour of a 'wild' animal. We found that at this youthful age our adopted pine marten did not like raspberry jam, which is a sure favourite with most adult pine martens, nor would he drink milk. His liquid intake was almost solely from his beaten-up egg.

Marty had very big paws on all his legs which seemed disproportionally large. Each paw was well cushioned and have five non-retractable claws. Body fur was dark brown, with this deepening to a darker brown on the legs and paws; dark brown bushy tail and handsome orange yellow frontlet. He had an alert, vivacious head with bright button eyes, and large rounded ears with cream-tipped rims, completed the picture of this handsome mustelid.

In going to sleep Marty went through a dog-like routine circling around two or three times when he would then do a sort of rolling somersault to lie on his back, forepaws clasped decorously over his chest, bushy tail curled around him. He slept very deeply, perhaps tired-out by his strenuous spells of activity.

His play began to exhibit new activities as he grew, such as spells of whirling in circles, incredibly fast, attempting to catch his own tail. His agility was breathtaking; he could run backwards, fast, all four legs synchronised, under perfect control. At other times Marty played a sledging game on the carpet, pointed muzzle ploughing along the carpet, all four legs spread out wide, sideways, from his body while progressing along on his stomach. Increasingly daring in his play Marty began to achieve new heights in the house. He disgraced himself one evening by extending his acrobatics to the scullery where, landing in an exuberant leap on the 'fridge top, he sent flying a container of milk. Investigating the resultant lake of milk, he paddled through this, then came bounding through to leave a trail of milky footprints on carpet and chairs. As I have said, his paws were large ones. The evening following this a plate of cakes went flying off this same elevation to smash on the floor and send a disconcerted marten accelerating back through to the living room. Only the great good nature and tolerance of my wife Margaret (did I ever say that she was brought up in an urban environment and that when we got married, she was apprehensive even of my Labrador dog?) prevented Marty from being banished outdoors for good. This would have been a bit premature, or so I persuaded myself. We compromised by trying to keep the scullery door closed thereafter when Marty was having his mad moments.

On a warm day in August Marty was put out permanently (*not* in disgrace) to his outdoor run. The time, I judged, was right; he was strong, fit and increasingly likely indoors to attempt stunts like swinging on the living room lighting. There was no protest from Marty, he even began carrying in bits of bedding on his second 'night out', in the shape of dry moss and heather, carried in his jaws to furnish his hutch. He was always glad to greet me each day as I went to feed him. We settled into the new routine, all seemed well.

Only six days later however I had a nasty fright for when I went out at 8 am to feed Marty it was to find him laying, stretched out, on his side, wheezing in distress, white foam around his muzzle. I picked him up to find that, as after his injection from the vet, he was quite dazed and unco-ordinated. I took him into the warmth of the living room at once and slowly, very slowly, he recovered. He was ravenously hungry and then, his appetite satisfied, he went to sleep, curled up on the bean-bag. We kept him in the house for a couple of days, then, since he seemed fully recovered, and the weather was warm, we put him out of doors again.

He took another of these 'turns' on 1 September; he was running around the run in an aimless, unco-ordinated manner and when picked up he, uncharacteristically, girned querulously. Indoors he came once more and this time he took longer to recover. For three days he refused all food, either sleeping or just lying about resting. Thankfully he then began to drink milk and that seemed enough to keep him alive. This time he was off colour for nearly a week and it was 7 September before I felt it safe to put him out in his run once more.

That, thank goodness, was the last occurrence of this mysterious affliction to which Marty seemed vulnerable, and for which at first I had blamed side-effects from the vet's antibiotic injection.

From then on, too, I began to enlarge Marty's diet and, since I wanted him to begin to accustom himself to eating 'prey' unaided, I ceased to cut food up for him. He coped well and our mutual 'education' course took another step forward. I began to try some fish in his diet, raw fish of course. Marty had saithe, haddock and cod and ate all three species. Strangely however, he did not like raw salmon. He had venison also and this too he accepted; I now know, in fact, that in winter, with prey scarce, a pine marten will eat from a dead deer carcase, which is that martens are not fastidious enough to turn up their nose at carrion when times are hard. I got a couple of unplucked wood pigeons for him and, after an initial and obviously irritating struggle with the soft feathers sticking around his jaws, he ably managed this addition to his bill of fare. He began to catch an occasional small bird in his run, evidence being left in a scattering of small feathers. Education progressing, Marty stayed with us until the following summer, when, again, probably under the influence of his first mating season, he opted, of his own freewill, to go in search of a mate. It may be that we saved his life; in entertainment and in interest value Marty repaid us a thousand-fold.

7. Wild and Free – Animals

The biggest thrill to me in the Torridon area, in seeing wildlife, was to find that sightings of otter were almost commonplace. For instance, Katherine MacKenzie, her car parked near the loch shore, came out from the Fasag shop with her messages one afternoon and flushed out, from below her car, two young nearly-full-grown, otters which fled back to the loch.

On another occasion Finlay MacKay, who lives in Inver Alligin and was for some years our postman, let his red setter dog out for a romp on the shore. To the incredulous astonishment of his wife, Mairi, and of Finlay himself, they saw an otter, a wild otter, emerge from the loch and for about fifteen minutes the dog and the otter had a sort of game of tag with each other. In and out of the sea, around a huge rock, slithering and sliding in the half-tide-exposed seaweed, they had a tremendous time in which no animosity whatever was displayed on either side. And Finlay had no film in his camera!

I found out, before we had been long in Torridon that a definite risk to the wild otters in the western coastal areas was that of drowning in a baited lobster pot. The pot being baited for lobster, the otter also was sometimes attracted by this bait, as, indeed were eels occasionally. Such was the case in one incident where a dead otter, an eel with an otter-bite out of the back of its head, yet still alive, and a lobster, were all found in one lobster pot. An otter, though such a superlative swimmer, is like humans in that it can only stay under-water so long; caught in a lobster pot in deep water it drowns. Remember also that otters, in 1969, were yet regarded as competitors for fish by most fishermen, although there was no excuse for this in our sea lochs. As an instance of this, one fisherman told me that he had run down and killed a couple of otters so engrossed in fighting, in the sea, that they had not heard his boat fast approaching them. Margaret beat me, by five years, in first seeing an otter 'fishing' for frogs in the wide and muddy ditch which bordered our driveway to the Mains and which eventually went into the river. Walking down to our Centre she heard a 'plop' from this ditch on a day in March. Curious, she went over to its edge, to see frogs, rudely interrupted in their spawning, scattering in panic, while a dark, whiskered muzzle, with, behind this, the dark line of back and tail of an otter, just breaking the surface, pursued them. This drain had been chock-a-block with mating frogs for a few days and the otter had found a rich hunting ground. Margaret has always had a nasty habit of upstaging me thus, in her sightings of wildlife; she had even actually seen the Loch Ness monster when we lived at Culachy.

My own turn was yet to come however, even though I had to wait five more years for this to happen. Again it was in March, on a cold grey morning after a slight fall of snow during the night. The frogs were not very active yet, since we had had a period of hard frost for some time. I was walking alongside the above-mentioned ditch, giving my terrier a pre-breakfast walk, while noting that the thin, wet snow was already beginning to thaw. I happened to glance casually up the long ditch which comes in from the hillside to intersect with the main ditch. There, part mud-wading, part swimming, was a black-looking, gleaming-wet otter. This ditch, and the main ditch, were swathed and mounded by quantities of bleached and dead mollinia grass, blown in by winter gales. The otter was stopping at intervals to search under these mounds, with its questing head and its shoulders at times completely out of sight. 'Frogs', I thought, and the otter was searching really thoroughly.

I froze, motionless, trying to look like a telegraph pole while feeling equally as conspicuous. I need not have worried; this otter's full attention was focused on food; a man and his dog were not going to put it off. When it got to the junction of the ditch it was searching with the roadside ditch, it glanced at us curiously, then, no more than a yard from us, turned up this main ditch. Where a drift of piled-high dead mollinia grass dammed the ditch completely the questing otter either poked and pushed its way through this, or, if this proved impossible, it clambered over the top.

Suddenly it pulled its head out from under a soaking heap of mollinia grass with a fat-looking frog, absolutely motionless, obviously deeply torpid with the cold, grasped in its jaws. The otter made no attempt to kill the immobile frog, which was upside down in its mouth; no, it just started eating. There was a very audible and rather nauseating scrunch of small bones as the otter ate its unconscious prey, beginning at the long hindlegs, with the frog now clutched between two fore-paws. The meal took only five minutes, then the head vanished into the otter's mouth, leaving only a tiny coil of spaghetti-like intestine on the ditch bank.

I decided I could safely leave this otter (which had immediately resumed its search for luckless frogs) to return to the house, put in my hard-to-restrain terrier and get my cameras.

I returned, panting, but I need not have rushed, for this amazingly unapprehensive otter was still intent on its frog-hunting. It was only about a further ten yards onwards as it probed below a pile of whitened mollinia. Slowly and thoroughly it probed and searched its way along the ditch, followed by my cautious steps. Now and then it searched so long, under grass, underwater, under mud, that it had to ease out to fill its lungs with air. I essayed a first photograph, a bit worried that the rather loud metallic click of the shutter would startle and put to flight my fearless otter. The otter heard it; it looked back over its shoulder at me, appraised my crouching form with jet-black, beady eyes, short, curled, light-coloured whiskers bristling, then turned back to the hunt.

In the ensuing seventeen years I was to see otters regularly, in the sea loch, in the river, crossing the road on a wet day, but never again did I ever have such an intimate glimpse into otter ways, or encounter such an enigma as that 'tame' wild otter. I saw otters catch crabs, crayfish, plaice, saithe, eels and other fish; I had them swimming and diving, eating crabs or fish on the seaweed-covered shore; I even had one curl up and go to sleep, while I watched, on a seaweed-surrounded rock.

The hill fox, a supremely vital and beautiful animal, is seen much less in the Highlands than in the English counties, or even in an English city. The truth is that in the Highlands, where hill sheep are present, or grouse are valued as a sporting resource, you will seldom find anyone to say a good word for the fox. Everyone's hand is set against it, the biggest achievement of which you can boast in a rural pub is of killing a fox, by shooting, snaring, or even by running one over with your car on the road.

This being so, the resourceful fox copes by being supremely elusive. Constant persecution strengthens a species; that the hill fox survives this constant persecution is testimony to its capacity to cope with this. I must say however that I have envied English friends who have seen, without rancour, from their bedoom windows, a fox breakfasting on windfall apples, pears or plums in their orchard. I doubt if some folk in the Highlands even realise that the fox will eat fruit. All of which is not to say that we should have an excessive number of foxes running all over the Highlands. Foxes are predatory animals but no one is more predatory by instinct, and often by practice, than humans. I am very fond of deer, of all species; I know, by personal experience, that fox, and eagle, prey on deer calves to some extent in the Highlands. Yet I do not detest either fox or eagle for killing what is natural prey to them. Grouse too are natural prey to fox and eagle; man has virtually replaced grouse in much of the West Highlands with sheep since this was thought to be advantageous. Largely as a result the fox (and occasionally the eagle), since they cannot take the now non-existent grouse, have been forced into killing lambs or into picking up already dead lambs. Man has destroyed the natural balance of nature.

Let me give an instance of how the fox is regarded in the Highlands. A farming tenant in Torridon had been loud in his complaints about the large number of rabbits 'eating "his" grass'. He came rushing to my door early one morning saying 'There's a fox in the field; come and shoot it.' Since this was in the lambing time and there were crofters' lambs in neighbouring fields to be safeguarded I took my rifle and managed to shoot this fox. I found that the lambs had been quite safe; still firmly clutched in the dead fox's jaws were two young rabbits.

Since the fox is so very firmly *persona non grata*, sights of this elusive animal on the hill are rare. Only once in all my life on the hill have I, in one day on the hill, seen two foxes going, unhurriedly, about their own affairs.

Just after first light on a morning in May I was away to the hill for a full day's prowl around. I was walking quietly along one side of a deep river gorge when I caught sight of a dead sheep across on the other side of the gorge. A handsome thick-set dog fox, in glowing red coat, bushy-tailed, with a big white tag at the end of this tail, was busily engaged in having a late breakfast. He was working away at the haunches, now and again forcefully tugging out wool, then feeding on the exposed meat. At times he just sat back on his haunches, busily chewing that meat which he had pulled off. At others I could sense his exasperation as he sat on his haunches to get rid of a tangle of wool which was clinging around his muzzle. He shook his muzzle from side to side energetically and scraped at it, with either forepaw in turn, trying to rid himself of this nuisance. I watched him for all of fifteen minutes, how long he had been breakfasting before I arrived I had no way of telling. Becoming replete at last the fox had a leisurely circuit of the dead ewe, pausing once to sniff at her head. Resuming his circle around the ewe he, all at once, made a lightning-swift sideways leap, cat-like in its easy agility, almost as if the ewe had suddenly come to life and lashed out at him. For a long moment he then stood, quizzically surveying the carcase, as if to make sure that it was definitely dead. He left then, pausing to anoint a small grass mound by the carcase before going out of sight. A small enough incident you may think, yet I savoured those few moments with one of our wariest native animals. The air was tangibly damp and almost without wind, yet I caught in my nostrils the faint elusive rankness of the fox. Perhaps on leaving the ewe, the fox himself had detected a faint whiff of human scent and had 'returned' the message. A polite salute between passers by? Or, much more likely, an impudent fox cocking a snook at this human who was impotent on the other side of the gorge.

Coming home, hours later, after a most enjoyable day on the hill, I saw a distant fox. Incredulous I got down and spied at it. A slim and rather tatty-of-coat vixen, this time slim but hardly svelte. She passed below me, a careworn air tangible about her. Her coat was a faded sandy-yellow and she had no proud white tag to her tail. Was she the mate of my breakfasting dog fox? Why was she a-prowl at this time of day, mid-afternoon? Had she lost her cubs and escaped herself, in the spring offensive on the fox dens? She passed below me and went out of sight. I could not remember when I'd seen two foxes in one day on the hill. How nice, it was, however (it was a Sunday) to be able to simply watch them, without rancour or overwhelming urge to destroy them.

Foxes, incidentally, are expert swimmers. A friend and colleague of mine once saw a fox swim across a fairly wide loch, carrying half a lamb in its jaws and, many years ago now, I saw one swim the Caledonian Canal. But then this ability to swim is a natural asset of our native animals. There are animals which one scarcely thinks of as swimmers such as mole, rabbit or wildcat. Yet all can, and do, swim when occasion demands it. I remember an occasion when winter floods had stranded sheep, away out on tiny islands of land surrounded by water. A friend, helping to rescue these (by boat, necessarily) was about to heave one aboard, which had been standing, chest deep, in floodwater, when he saw a mole perched on its dry back. Any port (or island, even if only of wool) in a storm. Mole and sheep, were duly rescued.

Adders, our only snake north of the Border, (and our only venomous one, though its bite is rarely fatal to humans) were scarce in the Torridon area, as indeed they were in the Fort Augustus of my childhood. Mind you this did not deter my mother, in her attempts to stop us using a path or area which she thought 'dangerous' to six-year-old infants with roving feet, from telling us that it was rife with adders. I'm afraid that it did not deter the aforesaid infants. I have seen only two adders, one in Invernessshire and one at Torridon, in more than forty years of prowling the hill. In both cases the adder concerned did its level best to avoid confrontation, by sliding away into cover. In the Inverness-shire incident I had the very devil of a job to prevent two colleagues, with me that day (and, like myself, armed with a shotgun) from blowing the adder's head off. Only when I averred that I was desperately keen to photograph a live adder did they accede to this queer request. In the other case, at Torridon, I had four children with me and we almost trod on an adder, which we then watched slither away, to our mutual interest, without danger to our mutual well being. Most of us, I'm afraid have a deep-rooted antipathy to snakes, indeed reptiles of any sort, as in my hillwalker's frog incident. This may have been valid in our distant past; it is not really valid now, yet the antipathy, and hostility, remains.

Perhaps the species of animal which gave me most pleasure at Torridon was the pine marten, both wild, or temporarily domesticated. These were much more regularly encountered at Torridon than in Inveness-shire in the 1970s. I had a request for aid one July from the new owner of a recently-acquired, long-unoccupied house which has been a church manse formerly. She had found that her loft, in this large rambling old house, was being used as a nursery by a family of pine martens. Would I come over and advise her? I did so. The house had been totally unoccupied for many years and I believe the pine marten, which has a penchant for unoccupied buildings, had been using this one for a year or two as a 'den'. The interior of the old house was being altered and various interior wood-lined walls were being breached, either to find water pipes or with a view to alteration. This being so, when I went to the succour of this lady in distress, she told me that the well-grown young martens had developed a habit of enjoying a romp in her bathroom, which they reached via the space between wood lining and outer stone wall. There can be few ladies in Scotland, indeed, in the world, who have enjoyed a hot bath while under scrutiny by two young pine martens, peering from a hole in the wall, unafraid and quite unabashed. I refer to the pine martens thus, but it held good for my informant also; it was a mutual admiration society. Luckily this was a redoubtable young lady who decided, (since I had told her the young would soon leave now) to leave her 'tenants' there, rather than have them forcibly evicted.

I did investigate the loft area for her and actually had one young marten in my hands. There was a certain distinctive, if faint, odour to be discerned but in such a big house it was liveable with. Also in the house was a large, grizzled wolfhound, of gentle disposition as most of this breed is nowadays, luckily, since their very appearance can be intimidating. Obviously, the martens were not intimidated.

The marten family was allowed house room until they left in early autumn. My redoubtable lady enjoyed many sightings of the mother returning about daybreak to the manse with prey for her young. Often this was of woodpigeon which was bulky for a pine marten but which was obviously caught while immobile when they were roosting in the darkness (quite safely, they probably imagined), high in a copse of trees. A large old sycamore tree opposite the house had a roomy, dry crotch in it and this the mother used as a daytime couch so as to 'escape' from the well-grown young after she had brought them breakfast.

I was later to 'acquire' my own truly wild pine marten. I had long suspected that young pine martens, in their youthful essays at attempting to provide for themselves in their first autumn, ate a prodigious amount of rowan berries. I had, for years, constantly found ping-pong ball-sized clusters of semi-digested and then regurgitated rowan berries, wrinkled in texture and orange, not vermilion orange but a pale orange, in colour. In this particular autumn my inquisitiveness led me to a large rowan tree situated on a hillside. One damp and consequently midge-tormented evening in early September I crouched, distinctly uncomfortably, behind a huge chunk of Torridonian rock, a grey cloud of apparently insatiable midges obscuring my vision. Ahead of me, squeezed into a cranny between two huge grey-pink rocks almost as tall as itself, was my rowan tree, so laden with berries as to appear more red then green-foliaged. To this rich feast I hoped a young pine marten would come. I also hoped that one would come before it got too dark for me to see it. It may seem incongruous that a meat-eater such as the pine-marten, should want to climb a rowan tree so as to eat rowan berries. However, pine martens which visit bird tables are demonstrably almost omnivorous and I knew that all pine martens relished fruit like blaeberries (their droppings were quite purple when blaeberries are available), and also raspberries and blackcurrants.

I had anointed myself well with anti-midge cream and without this I would have had to abandon my post. There was not a midge when I'd first arrived but within five minutes an opaque haze of midges surrounded my head. Then, at 7.20 pm midges were, at least temporarily, forgotten, for a flicker of dark fur whisked over one of the big rocks: a pine marten. From then onwards everything became worthwhile. It had flashed up the grey-barked trunk of the rowan in a trice to become hidden in the dense foliage above, only a periodic trembling of thinner twigs and some disturbance of the leaves giving its presence away. I found it hard to hold my binoculars steady, so excited was I, by this proximity to a wild pine marten. Once I saw the dark, bushy tail hang between two clusters of leaves, squirrel-like except for its dark colour. And then, quite unexpectedly, the pine marten emerged into full view on the outer canopy of the tree, surrounded by green foliage and vermilion berries. It seemed incredible that the slender branch tips of this outer canopy could sustain the weight of the marten, probably of about 3 lb. No doubt it was due to the way the lithe wee creature distributed its weight for the thin ends did not even seem to bend. Lovely in dark brown, glossy fur, this young marten was immaculate in its first winter coat and with a vivid orange-yellow frontlet. I became completely absorbed in watching the agile wee beastie, as much at home in the tree top as any red squirrel. So plentiful were the clustered berries that its method of eating was simply to cling on to the widely-spaced branches with all four paws and to reach upwards with its muzzle and nip off the berries within reach with glittering white teeth. Very occasionally one foreleg was used to reach over and pull a rich cluster within reach of the snipping teeth.

Suddenly a noisy, churring alarm from one of the family of ring ousels which were sharing the berries sent the marten scurrying down the tree. For a moment it paused at the base of the leafy cover, to peer around enquiringly, then it vanished into the rocks. Almost at once, as if realising that the ring ousel was just an alarmist, the marten reappeared and whisked up the tree again. The ring ousels indeed were in dangerous company for a cluster of dark feathers lay below the tree to testify that the marten relished bird flesh as well as rowan berries. I began to 'notice' the midges once more, initial excitement over. Indeed one could scarcely not 'notice' them; at one point I was fearful that the pine marten would detect my watching presence by the dark cloud of thwarted midges around my head. I re-anointed myself, smearing midges and midge cream in equal quantity over every exposed skin surface. Worst affected now were my stocking-covered legs; midges had discovered they could penetrate the open knit wool of these stockings, where I could not smear anti-midge cream.

All movement, except my involuntary twitching in midge torment, ceased for a short spell. Should I go home? Midge tormented, I was sorely tempted. I stayed and my reward came shortly. A dark sinuous shape shot up the exposed tree trunk, instantly followed by its twin. These were two young litter mates then, and both were now in the tree. I watched entranced as they re-emerged, running, vertically head down the tree trunk, unbelievably acrobatic. Time went by; I looked at my watch; no wonder the light was fading, it was 8 pm. The ring ousels were gone but I, reluctant to leave, lingered on. The martens chose to leave before me; I had a last glimpse of one of them running rapidly uphill, on a huge tilted slab of sandstone, thus avoiding the tangle of long heather which was all around the rocks. I have never seen martens move at slow speed, everything seems to have to be done at the double.

My delight was unbounded; I could not believe that I had been, at first attempt, so inconceivably lucky. In fact it was to be beginner's luck for though I saw one of the martens on subsequent occasions I never again saw two of them together. I put out supplementary food thereafter, on the huge rock by the rowan tree, and constructed a tiny hide, part-hidden by a huge rock. Rowans almost finished now, one of the martens appreciated bread spread with raspberry jam as a substitute. By dint of many evening visits, often in poor light, often midge-tormented, I got one or two photos from my hide. 'My' marten always materialised absolutely without warning, silently, not even a scrape of long claws on rock. It would pause for a second and with upward-held muzzle, quite obviously test the breeze. Then, satisfied, a couple of lithe, hooped bounds took it over to the bait. Watching it on a couple of occasions lick off all the jam from the bread, I next tried a spoonful of jam on its own. This proved very effective; so much of a sweet tooth had this marten that it even once carried off a piece of jam-soaked moss. Indeed, a 1 lb jar of raspberry jam, on which I had only a plastic bag as a lid, and which I'd hidden (safely I thought) high in a rock crevice, was found by the marten one night and the entire contents eaten. Better informed, I now used a jar with a secure lid, to find this one day, it's glass shattered, all jam gone, at the foot of its rock hidey-hole. Martens do love raspberry jam! I tried a hen's egg one evening; my fastidious young marten sniffed enquiringly at this strange object, then ignored it. I left it there, and the very next night the marten had a change of heart and took it away. I tried various varieties of fruit; brambles, blackcurrants, elderberries, apple slices and pear slices were all eaten; rose hips, holly berries (very similar in appearance to rowan berries) and tomatoes were ignored. Most meats, whether raw or cooked, were accepted, even haggis was eaten. Swiss roll, ginger cake and apple pie were other acceptable items.

I found that there were interlopers taking advantage of this largesse. Two voles lived dangerously for a while, then there was only one, while the skull and bones of the other I found in a marten dropping. A robin was another regular visitor. However the most intriguing visitor to the rocky banquet table was a tiny weasel, minute in size which, on one particular visit, tugged away a strip of venison scrap longer than itself. I continued to regularly feed this wild pine marten, 'my' marten, for years, until we left Torridon. I suspect that it was a female since it stayed for years in the same area. Incidentally, martens can attain thirteen to fourteen years of age, though I doubt if this age is attained often in the Highlands. As the years went by at Torridon, martens began to turn up increasingly in un-marten like places. Bird tables, in many areas, and with an infinity of diverse foods; a loft in an hotel, empty for the winter; another loft, in a disused stable, and strangest of all, on an auxiliary vessel used by the Navy, at Kyle of Lochalsh. A good friend of mine, the late Victor MacKay, a valued colleague who was National Trust for Scotland representative and factor for Balmacara, enlisted my help to solve this one, a nautical pine marten. He went with me to the marten-occupied vessel and we were shown down into the hold. On the way there I spotted a couple of unmistakable pine marten droppings, fresh, on the deck.

'We were at sea all day yesterday with some trainee personnel below deck, in the hold' we were told. One chap had come dashing up and asked what kind of furry animal we had down in the hold. He was obviously worried. Another crew member chimed in 'The pine martens have been aboard at least three of the boats, dashing around in the rigging and along the rails. They're a bit shy but not really frightened of humans. I've seen one scurrying along the edge of the pier, no fear at all of the deep water below. They have even been seen playing up and down on the radar mast.' I formed the impression that the regular crew were quite happy to have the distinction of a marten as stowaway on board. It was the other personnel which they had to carry from time to time who were concerned about 'the wee furry animal'. Perhaps they had reason; the marten, perhaps a wee bit stuck for prey in a ship's hold, on occasion shared, unasked and unseen, their packed lunches; half-eaten sandwiches and, once, a part-eaten apple had been found, explaining a missing lunch packet. I've no doubt that the marten went on shore leave when the auxiliary vessels tied up for the night. Indeed one of the security guards said a marten regularly raided the litter bin which was within sight of the guardroom.

I set a cage trap and, in due course, the marten obliged the Navy by entering this. Told of the capture, I drove over, collected the marten, innocently crouched at the end of the cage, and later, transferred it to quarters remote from a seafaring life.

Of other wildlife, we had no wild badgers at Torridon, with arable land at a minimum and hill ground so rocky and sterile it was not suitable for them. Nor did I ever see wildcats at Torridon itself, although I did, once or twice, while on Loch Maree side, in very early morning, (say 4 am, in summer) see a prowling wildcat. Common seals (these are smaller and rounder of head than the grey seals) but more often grey seals, came into Loch Torridon and occasionally afforded one a photo. Mountain hares were present but, since these are largely nocturnal, one only saw them in daylight if one came upon one among the rocks as it lay up, well camouflaged, for the day. I remember stepping round a huge rock one day and almost stepping upon one such hare. The eagles of the area seemed to find hares each nesting season, possibly after the hares had been disturbed into daytime motion by the same kind of surprise encounter as in this incident.

Stoats and weasels were both present, but, small and elusive as these were, one only saw them occasionally. Once however, I did find a litter of young stoats, eyes still unopened. I took a photo, then left them in peace. My best ever view of a weasel was as recorded earlier in this chapter, at my rock-feeding place for 'my' pine marten. I only once saw a stoat in full handsome winter ermine when I was rat-hunting down at the old farm steading.

8. Wild and Free – Birds

We have many exceedingly interesting birds in the remoter areas of the Highlands – golden eagle, merlin, ptarmigan, greenshank, the black-throated and red-throated divers, to name a few of these. Even in such distinguished company the dotterel ranks high as one of our rarest and most interesting birds.

The dotterel breeds in the Highlands on the long, green ridges found on some mountains at over 2500 feet, where short, sparse grasses and mosses predominate. Their main breeding range in Scotland is in the Cairngorms, but they breed and nest, in small numbers, on the eastern side of the Monadhliaths and occasionally in Ross-shire. The normal clutch is of three eggs and an intriguing fact about the dotterel is that it is the male which incubates the eggs and rears the young after they have hatched. The female however still condescends to lay the eggs.

I had long desired to see and possibly photograph the dotterel. My elation was tremendous when Mik, an interested friend, told me that he had, quite accidentally, stumbled across the nest of a dotterel in an area which must remain unnamed, except to say that it was in Ross-shire. He, sensing perhaps my scepticism, asked me across to see this nest and verify it for him. I did not need to be invited twice. On a rather misty day in June we set out to walk to the long green ridge, at around 3000 feet, where he had found the nest. The slopes of the hill became luxuriant, with a carpet of blaeberry, cowberry and cloudberry, as we trudged higher. Here and there were small patches of *betula nana*, the dwarf birch, hugging the surface, prostrate fashion. From a small gully ahead of us there ran a hind and her yearling, untidy of coat, the winter hair now loose and bleached in colour, not yet displaced by the thin rust-red of the summer coat. As we neared our first skyline, with short heather now beginning to give way to equally short grasses, a cock grouse, from his perch on a grey rock, rose, vociferous in protest, and whirred off over the ridge. Seconds later his mate appeared, furtive-looking, near to our feet and, instead of likewise taking wing, began the ages-old trick of a trailing, broken wing. Her brood of chicks, unseen up until now, rather spoilt her maternal decoy act by rising out of the short vegetation and legging it away, each one in a separate direction.

Higher still, and the vegetation was now a richly-patterned Persian carpet effect of lichens, mosses and grasses, with some sparse short heather. Another couple of hundred feet higher and a pair of ptarmigan materialised ahead, paused irresolutely for seconds, then took flight in the characteristic lovely white flutter of unfolded wings. The day had already been rewarding! Eventually we gained our objective, a long, featureless, green ridge over which rolled, intermittently, coils of thin, silvery-grey mist. It seemed to stretch endlessly ahead of us but it made beautifully effortless walking along its short, springy vegetation. Pessimistically I began to wonder how, or if, we were going to find the nest on that featureless ridge. Just as I began to sense a rather uncertain air about Mik, the dotterel rose off his eggs no more than a foot ahead. There, before our admiring eyes, lay the three typically plover-shape eggs, quietly elegant, brownish blotches on an olive-green background. They nestled in a small, shallow cup in the short grass and moss of the ridge, sufficiently set into the surface for the outline of the sitting bird to be barely, and certainly not obtrusively, visible above the short vegetation. The shallow cavity was very sparsely 'lined' with individual and now withering, blades of high ground grass, on which lay the eggs. I confess that I was speechless in my delight; up until then I had hardly dared to believe that I was going to actually witness a dotterel nesting in Ross-shire well away from the Cairngorm area.

The male dotterel, quite indisputable in his distinctive muted-in-tone plumage, meanwhile was circling wide around the nest, in a low, near-to-the-ground, crouching run, his tail feathers fanned wide, and depressed so as to trail along the vegetation. I obliged by being duped by his decoy act and went after him, trying for a photo with my long lens. My friend remained, motionless, where he was, chuckling at my mostly abortive attempts to photograph the dotterel. The bird, typically, kept my hopes alive by pausing every so often, then, just as I'd focussed, setting off on his, from his point of view, eminently successful decoy act again. All the time it was describing a wide semi-circle around where its eggs lay. Suddenly and unexpectedly the dotterel took wing and flew, in sharp-winged, flicking flight, straight back to the nest and, completely ignoring the rather substantial figure of Mik, still looming above the nest, settled unconcernedly on the eggs. The irony of the situation raised a wry grin from me and, I have no doubt, a belly laugh from Mik. There I was, having been led a fruitless perambulation around the area of the nest, and there was Mik, lacking a telephoto lens, now in perfect position to photograph the dotterel, not having moved a step.

My over-riding feeling, however, was of admiration for the dotterel who had successfully 'lured' me away and then returned onto his eggs. Trying not to notice the grin still on Mik's face, I edged slowly to within telephoto range of the placidly-sitting dotterel. My cautious approach was demonstrably needless; all that I had heard about the alleged tameness of the nesting dotterel was, it seemed, true. Our dotterel certainly co-operated beautifully, and sat there, placid and unworried, while I took more photographs. On that remote high ridge, only the overcast sky above us, we were on the top of our world, sharing the limitless solitude with the lone dotterel. Huge snow drifts lay in the northern gullies, visible from our high viewpoint. Reluctantly we left, the dotterel, quietly handsome and quite unafraid, still sitting on his nest. I was unashamedly garrulous on the way down; a long cherished ambition had at last been fulfilled; I had admired and photographed a nesting dotterel, in Ross-shire. I was drunk with elation; I had had a most memorable day, an unforgettable one, thanks to my friend Mik.

While I was at Torridon another quite remarkable first was that of the merlin which actually watched a pair of hoodies build a new nest and waited until this was completed, even to a nice warm lining of sheep's wool, before she and her mate usurped the nest for their own use. Lea, my eldest son, told me of this nest and, directed by him, I was able to find it and photograph it. The merlin is our smallest bird of prey in the Highlands and we had just one pair on Torridon; this handsome little raptor prefers moorland to mountain habitat and, while we had plenty of the latter at Torridon, moorland of any extent was scarce. Merlins do not build a nest for themselves; they either nest in the heather, hen harrier style, or they utilise the disused nest of a crow which had been built in the previous year. The pair had broken completely with all precedent in actually usurping the nest, the new nest, of the hoodies. There are not many species (nor many humans either) which can so conclusively best a pair of hoodies. It had taken repeated attacks, over a period of two days, to finally 'persuade' the hoodies that they had lost their nest. That established satisfactorily, the female merlin laid her four handsome, red-brown eggs on the deep, warm lining of the sheep's wool with which the hoodies had lined their former nest. I have seen hoodies regularly dive at and successfully harass the much larger buzzard, and, even more daringly, harass the golden eagle, but the hoodies in this instance had no answer to the piracy of the merlins.

I was able to watch the progress of this nest, and to photograph it, though it was poised high in the thinner branches of a slender birch tree. It made for rather unstable photography, one hand holding on to a branch in the swaying treetop while, with the other, I tried to focus and then to press the shutter release. Who was it who advised the use of a tripod for wildlife subjects?

On my second visit (12 June) there was absolutely no sign of the merlins as I climbed the tree and I was halfway up before the flutter of wings above my head told me that the merlin had left the nest. To my utter bemusement my head was on a level with the nest when the female, who had been sitting tight all the time, flew from the nest. The first bird I had heard had been the male and I had mistakenly taken his noise for the female leaving the nest. On the warm lining of the nest were two young merlins clad in white down; beside them lay one unhatched egg. Had a hoodie, in a lightning raid, taken the fourth egg, as part satisfaction for the loss of its nest? I left quickly, and as I descended the tree and departed the scene I saw the brown female merlin arrive to settle on her nest while, seconds later, the steel-blue male arrived in the neighbouring tree.

The next visit was four days later and I found both youngsters thriving, with one (a female probably) slightly the larger. A few small feathers, of a meadow pipit, I thought, were caught on one rim of the nest, the only indication of prey. On 1 July I made another visit. On this occasion both the adult birds circled above the top of the tree. The hitherto green leaves of the birch which were near to the nest were liberally whitewashed, as was the rim of the nest. As I reached nest level the smaller of the now well-feathered young fluttered off the nest in a premature essay at flight. I retrieved it and placed it back beside its nest mate, ambition quelled for the moment. Despite the untidiness of remnant wisps of down yet showing among their red-brown feathers they were already handsome little birds, with dark, liquid, expressive eyes seeming too large for their neat heads. There was no evidence of prey at all, indeed, apart from the accumulation of whitewash, the nest was clean and tidy. The young were not overtly or audibly aggressive, as both the young of the kestrel, and particularly, of the peregrine falcon would have been in similar circumstances. The two young flew safely a few days later; this was one contest of wits that the hoodies had lost.

On a positively bewitching winter's day, of soft, powdery, dry snow, a friend 'Badger' Walker and I, set out to look for ptarmigan on the high ridges. Ptarmigan have always held attraction for me; those bonny birds of the grouse family, speckled grey and white in summer, past masters in camouflage, and pure white in winter, phantoms of the snow then. Up on the ridge the bright sun was extracting pinpoints of sparkling light, glittering and gleaming, from the granular snow, virgin white, unmarked by any human foot. The snowy landscape was intensely beautiful and we both had our cameras at the ready. Tracks of deer, fox and ptarmigan were everywhere, the ptarmigan claw-prints especially numerous where any vegetation had been partially exposed by the wind. Here and there, small oval depressions, hollowed out of the snow, a few yellowish-brown droppings (rather like curved hazel catkins) at one end, showed where ptarmigan had spent the previous night, each snug in its snow hole. From these shallow hollows tracks radiated towards the potential feeding offered by the exposed vegetation, sparse enough in that Arctic world. Tantalisingly, however, there was no other overt sign of the feathered white ghosts of the snow. The myriad tracks aroused anticipation to fever pitch and we separated, each to cover a separate area of hill. Screes of sharp-edged rocks, sticking starkly through the white blanket, their cracks and crevices thickly cushioned by soft snow we explored, in case a snow-white ptarmigan might be lurking in their shelter. No luck! Well apart now, yet in sight of each other, we began to cross a wide expanse of crisp, part-frozen snow. Suddenly a 'patch' of 'snow' detached itself, pure white, wraithlike in the brilliance of sunlit snow, and began legging it, away from Badger and across my front. To see the snow-white bird with its bright, sealing-wax red eyebrow vivid on the white head, and, behind a brilliant, unclouded dark blue sky was the stuff of lyrics.

Even better was to come, for the cock ptarmigan, having legged a long way in the snow across my front, paused at a huge rock which was sticking, red-brown and sun-warmed, out of the snow and jumped onto it.

Cautiously I edged in closer, pausing whenever the head and neck of the ptarmigan elongated itself in puzzled enquiry. I took a series of photos as I edged nearer; the click of the camera shutter sounding sacrilegiously loud in that snowy solitude. Later on that same day in the evening sunlight we got pictures of a hen ptarmigan with her snugly-feathered feet also 'warming' on an exposed rock. There was no red eyebrow on the female bird so that her all white plumage was even better camouflage than that of the male.

In the sea-girt area of Torridon, where seawater takes such mighty bites out of the mountains, one got many rewarding sightings of the diver family both the black-throated diver and the slightly smaller red-throated diver. For much of the year both these divers spend their time around our coastal waters. In point of fact the only time that they come onto land is when they nest, inland, on freshwater lochs; they are built for a life on water, not on land.

For sheer elegance and sleek beauty of plumage these two divers are in a class all their own. Nor is this a flamboyant beauty (as in the case of the Slavonian grebe which also appeared on Loch Torridon occasionally) but a demure distinction, which gains, rather than loses, by its quiet blending of dove-greys, whites and blacks. The black-throated, to my eyes, is the more distinguished, with the supremely elegant dove-grey, velvet-textured head, striking wine-red eye, and long black vertical throat patch, bordered by alternate stripes of white and black. I defy anyone to distinguish individual feathers in the 'velvet' covering its neck and head, so close fitting is this velvet.

The deep red of the similarly vertical throat patch of the kindred species, the red-throat, is also lovely but it lacks the severe, clean-cut distinction of the black-throat's colour scheme, to my eyes, at any rate.

Both these divers are rare enough to be on the specially protected list in Britain. Few people today would imagine that such beautiful and interesting species like these two divers would ever be in danger from intentional persecution by humans yet, not so many years back, they were suffering from this, because they included very small trout in their diet. Bannerman, in his authoritative book *Birds of the British Isles* writes 'worst of all are those thoughtless and selfish fishers, mercifully not over-numerous who stamp on the diver's eggs when they come upon a nest in the remoter areas of the Highlands.' I was also reliably informed, in 1975, of how a certain 'gentleman' fisher desired his stalker to shoot a black-throated diver seen on one of his lochs. To his great credit the stalker refused, thus putting his further employment in some jeopardy.

In April each year both species begin to appear inland, on suitable breeding lochs. The black-throat has a preference for larger lochs, with islands, on which islands they may nest in preference to the loch's shores. The red-throat prefers much smaller lochans, at times only black, peaty pools, for this species flies back to the nearby sea to catch its food. Supremely elegant birds on the calm, still waters of a hill loch, the divers become creatures of comedy on land. Superlative swimmers they have webbed flippers rather than feet and these are set so far back on their bodies that they cannot walk properly on land. Their method of progression up from the water to, and from, their nest is a grotesque, breast-ploughing, sledging one, propelled by the pushing of the webbed feet. Their two long, narrow, dark olive green, black-splotched eggs are laid in an often damp, scantily-lined hollow, as close to the water's edge as possible. The two and fro ploughing progress of the diver's body eventually leaves a damp shallow furrow from water's edge to nest. In most cases this 'nest' is sited so close to the water that the sitting diver (which climbs onto the nest necessarily facing inland) reverses this position quickly so that she can dive literally straight off her eggs into the more familiar element of the water. A graceful clean-cut 'header' into the loch, in which it scarcely seems to ruffle the water with its torpedo-shaped body, and the diver is instantly below the surface, to rise, well out on the loch, out of danger. There, secure on the water, the diver will idle around until it judges it safe to come back onto the nest again. Both red-throat and black-throat, in a situation like this, may stage a mild distraction display, perhaps submerging elegant, velvet-grey heads below the surface, as if scanning the depths, while progressing, apparently headless, along the loch's surface. Or they may do a series of languid half-rolls on the surface, which displays the dazzling white of the underparts, not otherwise obvious. An even more spectacular distraction display, when those white underparts are even more prominently displayed, is when the diver rises vertically up, on tip-toe as it were, on the water, breast-on to the onlooker, thrashing its narrow wings furiously, before subsiding, to float tranquilly once more on the surface of the loch. I have heard the black-throat emit a muted melancholy, wailing cry at such times, as if imploring the source of the disturbance to go away. To me, however, the red-throat is by far the most garrulous of the two divers, more especially when it is on the wing. Time after time the inelegant hump-backed flight of a red-throated diver, returning to a nesting loch from the sea, is heralded by their utterly discordant, harsh, gabbling cry. I have also seen and heard the female red-throat, incubating, greet the return of her mate, heralded by his unmusical gabbling call with a low wailing note, head low to the water as she did so.

Situated as they are, their nests usually within feet of the loch edge, the eggs of the divers, indeed the sitting bird also, might well be thought to be very vulnerable to predation, by fox, for instance, yet I have never found any indication of this. Nor have I ever found trace of either diver at eyries of golden eagle or peregrine, and both of these predators hunt in diver-frequented areas. Thinking along these lines I have a foreboding that, one day, the alien mink (which by now is well established in many areas of the Highlands) may prove to be all too significant as a predator on the divers. The mink is, unfortunately, a very able swimmer. The diver's defence, apart from diving into the water, is by shrinking down low on the nest, head and neck extended flat to the ground. In this way the diver is simply a low grey hump without distinctive lines or silhouette, or indeed colour. I had a very interesting insight into the instinctive use of this immobility in a nesting red-throat one year. I had come on her at close range, suddenly and quite unexpectedly, as I walked around a lochan to photograph bog bean. She was sitting on her nest only a foot or so from the water and, it appeared to me that she could so easily have dived off her nest and into the loch. Instead, so imperceptibly that I could discern no movement, she lowered dove-

·grey head and neck downwards until it was extended, resting on the ground, flat out in front of her. Lying thus her wine-red eye stared unwinkingly up at me, from under her eyebrows. She maintained, unmoving, in this attitude while I hastily put my telephoto lens onto my camera and photographed her. It was a showery day and, on the tight-fitting cap of the grey velvet which was her head, the raindrops glistened, so that it looked bejewelled by them. Not a drop seemed to penetrate that velvet cap. I glanced back when I was fifty yards away; the extended flat shape of the nesting diver was well nigh invisible.

The great northern diver, (the diver called 'the loon' in America and Alaska, where it nests) I never did see until I came to Torridon. There however, on Loch Torridon, it appeared every winter. This is the largest of the divers (*gaviidae*) and reputedly a handsome bird. To my eyes, while it was a massive and a strikingly marked bird, it was not as beautiful and streamlined as either red-throat or black-throat. Nor is the white-billed diver, a fourth member of the species *gaviidae*, which turned up as a vagrant on Loch Torridon in the June of 1971. Because of the rarity of its occurrence in Britain it aroused much interest in dedicated birdwatchers. My admiration for the black-throated diver and the red-throated diver remained undiminished by my sightings of both of these kindred divers.

I should imagine everyone in Britain knows the success story of the re-establishment of the osprey as a breeding species in the Highlands, in the early 1950s. At one time ospreys bred on Loch Maree; I believe there were two pairs nesting on this long loch, which has also quite a few islands. Osgood MacKenzie, founder of Inverewe gardens (which is, of course, also under the guardianship of the National Trust for Scotland), had seen both osprey and sea eagle nesting in Wester Ross. The ospreys have not yet re-established themselves as nesting in this area but I am sure they will eventually do so. We have had single birds visiting the Torridon area; for instance a bird was observed diving at the salmon cages of a fish farm in Loch Torridon, being mobbed by gulls as it did so. Each time it dived, seeing the salmon swimming near the surface, it spotted the anti-predator net above the water, and pulled out, unharmed, before hitting this. Eventually, giving up; it flew off. I had the enormous thrill of accompanying Roy Dennis one summer when, with Colin, his assistant, he went to ring some young ospreys on a nest in Ross-shire. The nest was the usual bulky rather haphazard-looking structure; haphazard or not it had been strong enough, and firmly enough anchored to the very top of a half-dead Scots pine, to survive the winter's storms. A kestrel had attempted to usurp the nest for her own eggs a day or two before the osprey arrived back, only to be very summarily evicted when they did arrive. Both the parent ospreys were ringed birds, and from a previous nesting in the same area.

Roy and I watched while Colin climbed the tree, complete with rope and a small haversack. Colin lowered one bird in the haversack, Roy ringed it, measured it, noted down the statistics and the ringed bird was then lifted back up to the nest. There were three full-fledged young ospreys in the nest, red-brown of feathers, with orange-yellow eye and yellow legs and feet. The osprey has its claws so adapted that it can turn two to point forwards and two to point backwards, for greater ease in catching its fish. The young made no attempt at all to either fly or to scuttle away, when placed on the ground prior to being ringed, nor any attempt at aggression in striking with either beak or talons. There were, by their size, two female ospreys and one male in that particular brood.

The recording job finished, the young ospreys replaced in the nest, we left with the adult ospreys high above, circling the nest tree watching us depart. Every young osprey, in all the nests which were known, are ringed each year, before they leave the nest. The job is not always without hitch. While climbing a very tall, dead pine tree to one osprey nest, the RSPB official concerned was about sixty feet from the ground when he put his foot on a large, brittle, dead branch which contained a wasp's nest. Perched precariously above the ground, clinging on desperately with one hand, while using the other to swipe frantically at the wasps which surrounded him, he was not exactly enjoying the perils of ringing young ospreys at that particular moment.

The osprey began its re-establishment of its own volition in 1952, but has since been aided immeasurably by volunteer watchers. There are now more than forty known breeding pairs in the Highlands. Had each nest, as it was found, not been zealously guarded from the egg-stealers which nowadays infest the countryside each nesting season, the osprey might not have ever succeeded in its bid to return to Scotland.

There was a small heronry on Shieldaig island and there I visited one of the bulky nests, in mid-April one year. This nest was built on a stout side branch which extended out over the loch. An adult heron left the wide nest, with a derisive sounding harsh squawk as I climbed the tree. The nest was built entirely of interlaced thin and whippy branches, even to the shallow cup in the centre which held three recently-hatched young herons, a bristly, dingy-grey down covering them. They already had the javelin-beak of the heron and also that chill, yellow eye, devoid of warmth or expression, which is a feature of the adult heron. While I watched, two of the chicks arose, long stringy necks extended and had a juvenile sparring match, gripping at each other's beaks. Exhausted by this flurry of activity, they sank to a comatose grey heap on the nest, beside the third chick. The nest was very clean, no signs of prey, no white evacuations. Looking down into the shallow water below the tree I saw the pale blue eggshells of the hatched-off eggs gleaming there.

I made another visit towards the end of May that year. All the three young were well-fledged and very handsome now. In a period of about five weeks they had been transformed from the rather ugly grey and dowdy small chicks to a near facsimile of the elegance of the adult heron. One was inclined to be aggressive, darting its long, sharp-pointed beak at me. The two others cautiously sidled out onto slender twigs on the branches surrounding the nest, causing me to marvel at how these twigs sustained their weight without bending. Bulky as the birds looked, most of the bulk was probably accounted for by feathering, rather than flesh. The nest structure, and indeed some of the surrounding branches, was well white-washed now, while a quarter-pound trout, disgorged by one of the young herons, lay on an edge of the nest.

At my final visit, a week later, fragments of disgorged food passed my head to plop to the loch below – a warm welcome indeed. All three young herons, handsome now, perched on the branches surrounding the nest, while the disgorged remains of two quarter-pound trout lay on the nest. They were feeding well, well enough to discharge some of this at me, in protest, as I climbed the tree, a practice to which young herons are, nastily, addicted.

By 7 June all three herons were on the wing, flying strongly, though they did revisit the nest at intervals since the parents were still partly feeding them until they gained experience to feed themselves.

Like the golden eagle already written of, the young cuckoo arrived at the Mains entirely of its own volition. It first appeared near the house on 26 July one year where it took up stance on a fence. A great, sullen oaf of a bird, it dwarfed its meadow pipit foster-parent. She had to work so hard to keep pace with its insatiable appetite that I wondered when she had time to feed herself. You will gather that I did not regard the young cuckoo as an endearing character. A fluffed-out hulk of inanimate, dull-grey feathers, lifeless-looking until the arrival of the foster-parent with yet more food. Immobility then vanished, the beak gaped wide, to disclose the vivid orange-red chasm into which the food was poked, while, until food choked it off, a high-pitched whining squeak was emitted. The pipit came in every sixty seconds or so, usually with a small, wriggling worm, while I marvelled at the devotion she displayed to such a completely graceless youngster. The desire for food and yet more food seemed to be its sole motive in life; a driving, absolutely compelling need to get sufficient food to nourish it adequately for its soon-to-come long migration to Africa. To the human eye it was distinctly boorish in its behaviour, thrusting its beak petulantly at the pipit as soon as it had downed her offering, prodding its beak at her as if to say 'more, more quickly'. I christened it 'Oliver Twist' and wondered if Charles Dickens had ever studied the cuckoo.

On the shores of Loch Torridon, among the thousands of tide-washed pebbles, both oystercatcher and ringed plover nested. The ringed plover was masterly in its camouflage, both as to eggs, laid in a tiny hollow in the shingle, and in her own unobtrusiveness. The oystercatcher, on the other hand, was obviously flamboyant while sitting on her eggs; this she solved by doing as the peewit does; coming off her eggs, which are well camouflaged, and scuttling away, to rise, in flight, when she is well away from her eggs.

There was one laughable incident regarding a ringed plover's nest during our years at Torridon. A very nice and well-meaning lady when walking along the shore near to the Mains stumbled upon the four eggs of a ringed plover nestling in the shingle. She was rather perturbed, they looked so vulnerable and unprotected. An old, battered teapot lay, half-hidden in seaweed, a little way from the nest. The solution to the exposed situation of the nest came like a flash (some might rather say 'like a brainstorm'). Over to the teapot she ran, shook it free of seaweed and clapped it over the four eggs. Yes, that was much better; they were completely hidden! Indeed they were; it is the only time that I have known of the eggs of a ringed plover being almost infused. All ended well however; the lady reported what she had done, at my house. Due 'praise' being given, I then hurried down, when the lady had gone, and removed the guardian teapot. This well-protected ringed plover hatched its four eggs in due course.

It was on the pebbles of the shore, too, that I was walking, with my two terriers, one sunny morning. A ringed plover rose near my feet and scurried away, piping plaintively; a feathered 'low-flying jet' whistled by my ear, scooped up the ringed plover, without so much as a second's pause and a peregrine falcon was disappearing, fast, a last plaintive piping coming from its victim. The peregrine falcon is the most dramatic flyer of all our birds of prey and catches virtually all of its prey on the wing, that is, by flying it down. The eagle, by contrast, takes the majority of its prey from off the ground; a large heavy bird with a wing span of six to seven feet, the eagle has not got the dexterity of the peregrine in the air, nor does it need this dexterity. Each takes a very wide range of prey; in terms of size of prey obviously the eagle must be able to win, but in terms of variety I imagine the peregrine wins. I have recorded prey as tiny as a recently-hatched meadow pipit at an eagle eyrie and, at the other extreme as large as red deer calf or adult fox. At a peregrine eyrie I have also (once only) recorded a vole, and also a very tiny grouse chick while the largest prey I have recorded has been a lesser black-backed gull. However I have heard, from reliable friends, over the years, of a male blackcock being struck down by a peregrine and also of mallard duck. A friend of mine once had a meal of roast duck after a peregrine had put this into a rush-grown ditch. I have also had a meal provided by a peregrine – in the shape of a grouse one winter. In this case the peregrine flew down the hurtling grouse, bound to it, grip-ping the leading wing edges close to the body, and, riding pick-a-back, reached forward and severed the spinal cord at the back of the victim's head. What did have me non-plussed was the fact that the peregrine planed down, carrying the stricken grouse, landed with it, stood over it for a moment, then flew off, abandoning this kill. Returning, in evening, the grouse was still there and I confess I confiscated it.

Like the kestrel and the merlin the peregrine falcon does not build a nest. One spring, when visiting an eagle eyrie at Torridon, I found four very handsome red-brown eggs, eggs of the peregrine, lying in the centre of the huge nest. I left quickly and returned in mid-May by which time I judged the eggs should be hatched. Both male and female peregrine came from the cliff site, the female from off the nest, and the male from his guard rock near the cliff top, shrieking abuse, harsh, guttural shrieking which was quite intimidating and distinctly unmusical. White and grey feathers spoke of ptarmigan as recent prey. Two of the four eggs had hatched and two tiny morsels of white down, through which the skin showed pink, lay beside the two eggs. Fragile in appearance there was no hint yet of the rapacious pirates of the skies into which they would develop. Their juvenile beaks were pink, mother-of-pearl looking; their eyes, unopened, were 'blinkered' by opaque, blue-grey bulges of mem-braneous skin, reminiscent of the falconer's hood. I left, and later, on the opposite side of the narrow glen, lay down in the long heather and spied back at the eyrie. The falcon was on her eyrie again; through my 20X telescope she seemed to look straight at me with large, liquid, black-looking inscrutable eyes. She was a striking bird – slate-blue head, dark moustaches streaking her cheeks, pale upper breast with the veriest hint of salmon pink colouring, lower breast flecked with dark bars. Her legs I knew to be bright crayon-yellow, plusfoured in feathers, black-barred, to mid-leg, and ending in the long prehensile talons of the feet with which she binds to her prey.

On my next visit, on 25 May, all four eggs were hatched, with the four young, in varying sizes, huddled in the nest centre. I had taken a small spring balance with me to weigh the young and this I quickly did, then left, to the discordant outrage of both parents. The largest young one weighed twelve ounces, the second one eleven ounces, while the two last to hatch were much lighter in weight, three ounces and two ounces respectively. At this early age they did not need much feeding and only the remains of a grouse lay on the eyrie. The female, larger and bolder than the male, stooped very close to my head as I clambered down and away from the eyrie, and nearly deafened me by her screeching. I passed a pair of greenshank on my way home; they had dangerous neighbours.

My next visit, in early June, was after a night of heavy rain. The eyrie under its deep, sheltering overhang, was absolutely dry, as were the young peregrines who were huddled together for warmth, as usual. The three biggest began a juvenile harsh screeching in a fairly creditable imitation of their parents, meantime leaning backwards and striking out with their already large taloned feet. The nest was now a tawdry carpet of 'whitewash' and pieces of prey. I identified black-headed gull, grouse, ptarmigan and greenshank. I re-weighed the fast-growing young; the first hatched was now two pounds two ounces, the second was two pounds while the third was only one pound six ounces and the smallest, the runt of the clutch was only one pound two ounces. The largest had tiny cheek-patches of orange-brown feathering now, and their beaks, at three weeks after hatching, were blue-grey instead of pink. Their bodies were still mainly clad in white down; this down, however, was a dingy grey-white now, with patches of it bloodstained. A week later, when I arrived, the female greeted me with her usual burst of harsh, guttural, peregrine invective. At first I could only see three of the young peregrines. Had the 'runt' died? No, it was one of the three; that I could see by its 'runty' size. Then as I looked along the ledge I saw the fourth young one, clutching the almost flesh-less framework of a herring gull, obviously its own private hoard, dragged away from the others. All four now combined in setting up a deafening chorus of harsh, ear-grating screeching. On the nest's centre lay the part-eaten carcase of a lesser blackbacked gull, white head, white body, gruesomely bloodstained, vicious beak, bright yellow with its red dot, and bright yellow webbed feet. The weights of the young peregrines were exactly as they had been a week ago, except that number three had gained two ounces. The prey identified, (and part of it was unidentifiable) consisted of the two gulls already mentioned, two sandpipers, one ring ousel, one mistle thrush, one shearwater, three rooks, two grouse and a razorbill. There was also a leg ring from a racing pigeon. Did I say that the peregrine was a very efficient provider of prey for its young? By pure coincidence, I was also visiting the eyrie of a golden eagle regularly that year, and I can assure you that the show of prey visible there was much less, both in quantity and variety, than that at the peregrine eyrie.

It was eleven days before I got back to the peregrine eyrie. On this visit, the four young, all well-feathered now, were dining, en masse, on prey obviously just freshly brought in. This prey was quite unidentifiable, hidden by the heaving mass of red-brown-feathered young peregrines. I was a little surprised to see that the 'runt' was competing with the best of them. Perhaps it was going to make it. Prey remains covered the entire expanse of the eyrie in a squalid tangle of bones, feet, wings and loose feathers. I could see no seabird remains this time, but there were identifiable four grouse, one ptarmigan, one crow, one greenshank, one sandpiper and the skull of a ring ousel. I should have explained that I cleared the eyrie of all prey remains after I made each visit so that I could identify fresh prey on a later visit.

My last visit was near the end of June. There were the three larger young, and, sadly, further along the ledge, lay the body of the 'runt'. It had not made it, after all. The three surviving birds were now all handsome, in red-brown feathers, through which vagrant wisps of thread-like down poked. Their voices were stronger also and I was almost deafened by the incredibly harsh screeching. First one, then the second of the two larger birds, females obviously, sprang from the ledge of the eyrie and flew in a flicker of narrow wings, from the cliff face. The third, probably a male, shrank as far into a rock crevice as it could and there I left him. He would be gone in a couple of days. The two which had flown had taken forty days from hatching to flying, amply nourished along the way by their parents. Gamebirds had, incidentally, rarely featured in the prey brought into the eyrie.

D. F. TUNSTALL
Fellside
Barwise Road, Arlecdon
Frizington, Cumbria
CA26 3XD
Lamplugh (0946) 861051

9. *Tame Red Deer*

I have enjoyed keeping deer nearly all my life but then I have had the space and the opportunity to do this in a responsible manner. This has given me immense pleasure (and some frustration) but I must sound a warning here. Tamed female deer, hand-reared from birth, are usually dependable and predictable; tamed male deer, whether red or roe, even when also hand-reared from a very early age, are the very opposite and I for one, would never recommend that anyone should attempt to keep a male deer as a pet. It does not matter how affectionately these are treated, as undeniably attractive youngsters or how seemingly affectionate they appear to be in return. The male's rutting instincts and mating patterns inevitably render him dangerously aggressive just as soon as he attains an age when he is influenced by this overriding instinct. Moreover, as he grows older, he grows increasingly familiar with his human owner and it is a true saying that familiarity can breed contempt; your male deer seems to lose all the instinctive fear of humans, which, centuries old, inhibits all wild species, including deer. A soft-skinned, soft-muscled human, relatively slow in reflexes, is comparatively helpless, unless armed in some way, against a red stag (or even a roe buck) which may be inflamed, beyond all reason, by the compelling fever of the rut. The age at which a red stag, hand-reared, becomes unpredictable, and hence dangerous, I would put at three to four years old, while, with a roe buck, at one to two years of age.

I have always stressed the point that I have never felt any apprehension whatever, even at the rutting season on the hill. The slightest suspicion of the presence of a human being, the veriest glimpse, whisper of sound, or whiff of scent and your intimidatingly belligerent stag is running for cover. Your hand-reared stag, instincts blunted, knowledge of humans sharpened by his hand-rearing, is quite another matter. It was, in fact, my own enclosed and 'domesticated' stag which almost ensured me of a lasting niche in red deer literature. It was in October; my stag had been roaring for days, and he was giving his two hinds very little peace or any time at all for contemplation. I walked alongside the deer fence which enclosed the stag (too close, to it, as I realised in due course). He, wrought up in his rutting frenzy, came across, roaring, to pace parallel to me, but, of course, inside the fence. I should have recognised that slow-motion, stiff-legged gait, menace implicit in every slow pace, every hair on the stag's body and bristling shaggy neck stiffly erect. I had witnessed this often enough on the hill, at the rutting time, as two big stags paced thus, side by side, sizing each other up, as a prelude to combat or to a strategic withdrawal on the part of one of them. Foolishly, I did not make the connection and even more foolishly I essayed a bogus roar, in reply to the roaring of my stag. As the basis of my disregard, I should explain that I just did not see myself as a male competitor for the hinds of my stag. Not, that is, until this was brought, forcibly, to my notice. There was an incredibly fast blur of movement from within the fence and something hard smote me with irresisitible force about my lower left ribs, accompanied by a twanging and creaking of the sorely-stressed fence. I was unceremoniously thrown completely off my feet and, completely airborne, landed in a confused and untidy heap a full twelve yards away. Strangely, my initial reaction was one of bewildered rancour and aggravation; how dare my stag do this to me. This was very quickly followed by a feeling of absolute thankfulness, as I saw the maddened stag, thrown back at first by the elasticity of the wire as this reacted to his jet-propelled rush, advancing again, thrusting powerfully his long and lethal brow points through the mesh of the fence, in a determined attempt at finishing off his fallen 'adversary'. I have little doubt that he could have done this, before I had even regained my feet, had it not been for the lifesaving fence. Believe me, 'tame' stags should always be regarded as unpredictable. This particular stag I, thereafter, always kept enclosed and never again did I tempt fate at the rutting season. Some years later he died, his teeth worn out, without ever getting another chance to register a victory over one of the human race.

A son of his was born to my oldest hind. She was a veteran hind by then and she had very little milk for her stag calf. Her calf did have that vital first drink from the hind, which contains the invaluable colostrum; thereafter, since the mother then opted out, I hand-reared him, successfully, without any foreboding of what was to ensue, years later. Indeed he showed, in his early years, no sign of the underlying aggressiveness which had distinguished his sire. I often took children of visiting friends to see him in the summer when his antlers were in velvet. The children used to feed sprays of fresh green leaves to him while parental cameras went 'click' at this wondrous happening. My 'tame' deer gave immense pleasure to a multitude of visitors in our years at Torridon; it was a novel experience to most visitors, to be allowed to go so close to deer, in most instances the dependable hinds, and to be allowed to feel their wiry coat, while a moist muzzle quested for a titbit. One strict rule which I always enforced was that no one was to go into a deer enclosure unless I was with them.

The character of this bottle-fed stag altered dramatically and drastically when, aged five and with his sire dead, he took charge of the hinds. I returned from the hill on an evening in October to find my sorely-tried wife in a fearful state of anxiety. The stag, fearsomely black and dripping wet from a fresh wallow in the black, peaty hole available to my deer in their enclosure, had literally gone beserk. Neither of the two hinds, which were in with him were in season and he very obviously resented this. The younger hind, only a yearling, he chased round and round, in and out of the communicating enclosures, ceaselessly chivvying her until she lay down, exhausted. He then took his spleen out on her, as she lay unresponsive, striving to 'nudge' her to her feet to resume the chasing 'game'. His thrusts, with his brow points, grew increasingly more violent and frenzied until, in the end, her gored her so severely that, after I had returned home and carried her to safety, she died in the night. Her body had massive injuries through the multiple stab wounds of that completely unreasoned attack.

Unsatisfied, and in a state of frenzy which defies description, the stag then transferred his aspirations to the other hind, a four-year-old one, who had her calf of the year at foot. Incredibly, such is the power and muscle of the enlarged neck of a rutting stag, he succeeded in forking her completely off her feet and hurling her so violently into the top netting of the deer fence that she somersaulted over the top and landed in the field outside. This, I am quite sure, saved her life! Unpredictably, the stag, inflamed as he was, did not molest the calf in any way.

I had to make a decision then and there; I did not want to have to shoot the stag unless all else failed. I chose to segregate him in one division of the enclosure, tempted him into this with a bucket of deer nuts, then closed him in. By morning, though now apparently calmed down somewhat, he had demolished, with his antlers, sufficient of the dividing netting to get back into the main enclosure. I somehow contrived to get him shut in again though he was once more approaching his former frenzied state. This time I strengthened the dividing wire so that I was certain he could not again force his way through. I should really have known better; by then he was so utterly possessed that I had never seen any wild stag so demoniac in appearance or activated by such a frenzied restlessness. He paused neither to eat nor to rest but incessantly patrolled the fence, probing for a weak spot.

By the next morning he had found one and was in the main enclosure once more. Fearing for the safety of the hind and calf, and indeed, by now, a trifle apprehensive for my own safety, I took my rifle in with me this time while I tried once more to segregate him. As soon as I ventured in to the enclosure he advanced on me, wild, shaggy of mane, madness in his eyes. Regretfully, I put a bullet in his neck, at point blank range. I had used up all my alternatives and had signally failed.

You will have gathered by now that looking after tame deer is not without its disadvantages, yet, knowing what this entails, as I do now, I would do it all again. The anxieties can be great, the inevitable heartbreaks, (as a favourite hind dying while trying to give birth to her first calf, or the death of the young hind by the frenzied behaviour of the stag) even greater. The rewards however can also be great and, in the long run, this was what mattered.

On a day in June at Torridon I was confronted with another problem; this time it was not one of a death-dealing stag but one of a life-giving hind. I had decided to remain at home that day and it proved very fortunate that I had not opted instead for a day on the hill. A hind was due to give birth soon, the same hind which had, in a previous year, been thrown out over the deer fence by the stag. She also happened to be a favourite of mine because I had had to hand-rear her and she was exceedingly tame and affectionate. I had called her Beauty, the same name that I had given to the very first calf I had ever hand-reared, at Culachy, above Fort Augustus.

During that morning I checked on her periodically but it was not until 2 pm that she lay down and showed signs of starting to give birth. This would be her third calf and she had had no problems in giving birth to her two previous calves. I checked on her, then left her in peace to give birth in her own time.

It was not until 5.30 pm that I returned, expecting to find that the new calf had arrived and that all was well. I could hardly have made a more incorrect assumption. The calf had not arrived, or not altogether, the hind was in difficulty giving birth to it. The head of the calf protruded, under the tail of the hind, as did half the length of one foreleg, with a much shorter length of the other foreleg. The extruding head appeared completely lifeless, matted, black, wet-plastered hair, eyes closed, nostrils apparently partially plugged with mud, the limp tongue protruding. To be truthful I was convinced that the calf was dead and that unless I did something I would have a dead hind also. My first problem was to get my hind to allow me to help her; tame enough to let me fondle her head, scratch her ears, take a tit-bit from my fingers, she nevertheless shied away decisively when I tried to catch her calf's head so that I could try to help her. I watched her for a while, lying down, straining down vigorously, her head turned back towards the protruding lifeless-looking head of her calf as if exhorting it to help her. A while she lay, then sprung to her feet as if stung by a hornet, while a visible ripple shuddered its way down her flanks. Hunched up, she stood then, visibly straining, but unavailingly. I went for a length of soft-fibred ripe I had in the garage; I would have to try and secure her somehow or I was going to be quite unable to help her. I was already reconciled to the loss of the calf.

Back in the pen, a bucket of deer nuts lured Beauty to me; head down, she munched at the nuts and I got a loop of the rope around her neck quite easily. All suspicion of easiness ended there and then; the world thereafter exploded around me. There ensued a type of rodeo, with my hind acting the part of the roped steer, as spectacular as anything I had ever read of, or watched on the screen, portraying a Wild West rodeo. It proved to be, also, quite as brutally exhausting as ever I had imagined a rodeo to be. I was being pulled all over the length and breadth of the large tree-grown grassy enclosure, on my knees, or sliding on my rump, but, mostly, I admit, at full length, prone on my stomach over the grass, by a pulling, plunging, writhing, bucking, hoof-lashing hind of about 200 pounds weight who, her travail of calving forgotten, seemed to have developed inexhaustible energy from some deep and quite untapped reserve. I lost count of the times I was dragged, at full length, while desperately trying to halt the hind's progress; my wrists, around which I had looped the rope now both muddy and bloody. Exercise may be good for expectant mothers but I had now very grave misgivings about the effect this type of exercise was going to have on my hind. I already was experiencing its effects on me.

At last, and it took an unbelievable time, (during which the notion, incongruously, struck me that it was a good job I hadn't been wearing a wristwatch) the rampant hind tired sufficiently of towing my 140 pounds of weight about and slowed up enough for me to snub the rope around a tree. We lay there, side by side, red deer hind and male 'midwife', for minutes, gasping for breath, while I mouthed, with what breath I managed to summon up, soothing murmurs to the distraught hind.

Reluctantly I exhorted myself to the task I dreaded, and the reason why hind and human were laying exhausted side by side – I would have to try and take away the dead calf. Amateur midwife that I was, I tried first simply a sustained pull at the head of the limp calf. To no avail; I remembered, in my far off boyhood, that I had once seen giraffe-necked women at a Commonwealth Exhibition staged in the Glasgow of the late 1930s. I figured that, if I continued the pulling I would end up with a giraffe-necked deer calf. Racking my brains for recollection of how I had seen a lambing shepherd act in a similar predicament with a blackface ewe, I then tried to ease one most appallingly unhygienic and bloody hand in alongside the stiffly sticking out foreleg, with its black, yellow-edged hoof. My hand went in surprisingly easily and from then on it was fundamental common-sense. I felt along the length of the shorter of the forelegs to try and find how and where it was wedged. A gentle easing out where it was tightly wedged, a simultaneous bearing down by the now co-operative Beauty, and the entire, repulsive, slimy, glistening-wet length of the 'dead' calf 'flowed' out in a gush of newly-released foetal water. Neglecting even to look at the 'dead' calf I at once released the hind from her securing rope, while, tired and beginning to ache as I was, blissful relief washed over me. I had saved the hind and thought we had lost her calf. Nor did Beauty as much as spare a glance at the calf, lying still half-enclosed in the membraneous envelope of its birth. Her trust in me completely ruptured by my 'betrayal' she ran to another section of the enclosure. I looked wearily at the 'dead' calf, to be literally galvanised into incredulous action again; the 'dead' calf was stirring feebly, its hitherto closed eyes were open, its legs weakly straddling; a reincarnation was taking place under my amazed eyes.

I 'cleaned' the calf myself, clearing its nostrils and mouth of all the clinging mud first. Then, using my hands where the mother hind would have used her mouth, her teeth, and her tongue, I got rid of all the silvery-grey translucent 'skin' which still partially clothed the calf. The new-born calf was scarcely a thing of beauty with its slimy and damp-domed head, huge sail-like ears, likewise damp, and black-looking, slimy body with no dapples discernible. Stilt-like legs culminating in dark hooves, soft and plastic to the touch, which were edged with yellow-white, completed its appearance. It seemed unbelievable that, within the next twelve hours, this ugly 'object' would be transformed into the dappled loveliness of the young deer calf. My midwidery finished, I carried the calf, a female, over to where the exhausted hind waited. Laying her down, I retreated quietly; the black-looking calf lay there, moving only its head, weakly, as if unsure of its welcome. Once more that evening relief washed over me as I watched the hind slowly approach, sniff at the prone form of her calf, and begin eagerly to lick it clean of the birth 'slime'. Later that June evening, a trifle refreshed but beginning to find new bruises and abrasions whenever I moved, I watched the miracle calf have its first life-giving drink from the now solicitous hind's udder. I weighed her afterwards; she weighed eighteen pounds, a good weight for a female calf. And so I had expanded my duties at Torridon to that of a cervine midwife; it would all have been so simple if the hind, tame as she was, had trusted me to help her.

The female calf progressed well; I kept a proprietary eye on her, weighing her at fourteen days old when she weighed twenty-eight pounds, and again at twenty-one days old when she weighed thirty-two pounds. By then she was a lively, healthy deer calf and was now accompanying the hind everywhere. My apprentice 'midwifery' had, thankfully, done no harm.

Since I had had to put down my rut-inflamed, berserk stag there was a period when I had no stag. Many visitors of that stagless era expressed surprise that I had not got a stag, yet, due to the enterprise of Beauty I had a dappled calf in the enclosure each June. The explanation is simple. Each rutting season Beauty exhibited sufficient signs by which I knew that she was beginning to harbour certain very basic notions which could not be satisfied in the wholly female society in her enclosure. This being so I would open the gate and let her out when she purposefully set out to find her own version of Prince Charming. This aroused amazement in all my listeners, perhaps not so much that I had set her free to go to the hill but that I was completely confident that she would, in her own time and entirely of her own volition, return back to the Mains. To me this confident expectation was not remarkable. I knew that red deer hinds are strongly hefted to their birth area and I knew that all red deer are very much animals ruled by habit. I knew also that deer always remember an area where the feeding is good; I knew that Beauty was attached to a certain degree to me, and to her previous offspring who were still in the enclosure.

I once had a roe deer doe who had complete liberty around my house and who had left to go to the hill, about the time when her breeding instincts began to assert themselves. She was gone for ten months, by which time I had given her up for good. No such thing; she returned 'home' to have her first fawn where she herself had been reared. My confident expectations, then, were not so ill-founded. Not that I did not worry; the worst risk really was that she would have to cross the main road, and she was not used to traffic. The risk of an opportunist poacher, touring the lonely glen road in the hours of darkness, with vehicle equipped with spotlight and firearm, was also there. So even though I knew that these risks were slight I tended to grow more anxious as each day went by and Beauty had not returned. Each time that she went to the hill however, so did I always find her, a satisfied wanderer, honeymoon over, back at the gate one morning, ready to return to her enclosure. This denouement was always enormously satisfying; one got the comforting impression that the hind was happy with her life, else why did she come back to its slight restrictions instead of remaining on the hill?

I should perhaps explain that such an 'arrangement' does not always work so well. While we lived at Culachy I had a hind which I had also hand-reared and who was in no way confined. As with my roe doe my hind simply lived beside the house and indeed quite often invaded this house. Apple tart which Margaret baked and would, pre-hind days, leave out on the windowsill to cool, was a sure lure. One Christmas time she came in, the front door being left hospitably open, and polished off a bowl of tangerines before she was ejected.

To come to the point however, there was no stag available at the house on Culachy and, one day in late autumn, being aware of the nuptial signs, I decided to lead this hind (also dubbed Beauty) to a shallow coire in the hill behind the house, where I had seen stags that morning. She would follow me as faithfully as any well-trained dog, (at times, to my embarrassment) and so away we went, in convoy, to find the stags. I located the stags, lying chewing their cud, their rutting aspirations more or less dormant now, the winter not far ahead. My hind followed my crouching form, then my crawling progression and finally my tortuous stomach-hugging-the-heather approach. The idea was to get her close enough to these antlered 'cavaliers', myself unseen, to let her see them and to be herself also seen. I got close enough unseen but there the programme irrevocably broke down. These stags, their rut over now, were no ardent suitors while my hind probably did not even realise that they were of her own species. Desperately I tried, while lying prone behind a rock, to convey to my urgently-in-need-of-male-company hind that there, within each reach, were plenty of suitors but she just would not leave my prone form. She was apparently convinced that she also was a human and she saw no attraction in these antlered beasts. Nor apparently did they see any attraction in my buxom young hind.

Things reached an apparent impasse; my hind, breathing heavily, now endeavoured by dint of sundry shrewdly directed blows of one foreleg, to get my prone figure to its feet, presumably to prove my masculinity. Loathe to prove my incompetence in this, I lay on, gritting my teeth, hard. At long last she gave me up and began to do what I had hoped for; she grazed away from me, towards the cosily recumbent stags. I sighed in relief; matters were progressing.

Absolute disaster was to follow; incredulously, I watched her thread her way through that bunch of anything but virile stags, not one of which even ceased their chewing of the cud. Maybe it was going to take some time for interest to be aroused; I wriggled back out, slowly and cautiously, and made for home, leaving, I thought, my hind in congenial company. As I reached my door an all too familiar stertorous, heavy breathing sounded at the back of my head. My hind was still convinced that I was the solution to her marital difficulties; she had followed me back. Have you ever been assiduously courted by a hard-breathing, body-rubbing red deer hind? Flattering it may be, satisfactory it is not, for either party.

That very night, as darkness fell, my hind, disappointed in one male of the genus homo, decided to carry her quest further afield. Still breathing throatily, wagging her short tail vigorously, my single-minded hind set off, not to the hill, but to the village, about three miles distant. I was quite unaware of this until I received, by 'phone, an urgent request to come and remove my hind before she was arrested for importuning the male members of the local British Legion club in Fort Augustus. I did so, poste haste, and, thankfully, by next day, my courting hind was over her fever and more interested in illicit apple pie than in importuning astounded British Legion members as they went for a quiet drink.

However, my Torridon hind knew where she was going and also was quite clear what an eligible suitor should look like. When she was let out, on the evening of 19 October, she purposefully set out for the hill and returned, mission fulfilled, some ten days later, palpably satisfied. One visualised a bemused stag trying to head her back into his hinds as Beauty, now no longer wishing to know him, headed, determinedly, for home. Food and shelter, after all, were more important than one moment of cervine 'bliss'!

The following June Beauty proved that she had had every reason to appear satisfied when she gave birth to a strong, female calf, of nineteen pounds weight, to the sound of that self-satisfied, bovine-like bellow, which marks the calving hind in the Highlands, in the calf-bearing month of June.

Precedent now established, Beauty took her way to the hill in the following October also, this time for only three days. The result of the union which demonstrably occurred, with a stag which will forever remain incognito, was to come in the summer after her successful foray. I remained quite unaware that the event which was to occur then was to be an unforgettable landmark in all my years of dealing with deer. I had always believed that, on very, very rare occasions indeed, a red deer hind in the Highlands will give birth to, and subsequently rear, twins.

What followed was the most incredible event which has ever happened to me in a not entirely uneventful life, a life shared, enjoyably, with deer. It took place on 14 July 1982 when Beauty gave birth to twins and I was present to record their birth. Even to this day, ten years later I regard it as incredible that it should have happened to me, to have my hind give birth to twins, neither 'arranged', nor anticipated, their very sire completely unknown. Moreover, I had the knowledge to realise how how rare an event this was, hence the incentive to record the details of the birth, and I also had the necessary equipment to record this photographically. The snag (and I hardly regarded it as such) was that it was a vile day for weather, dull, darkened by heavy clouds, and raining heavily. This weather in fact was what had kept me from the hill on that day; the balance, therefore, was on the credit side. Beauty's twins, born on that July day in 1982, made only the second eye-witness record account ever of a twin birth in red deer in Britain.

For almost four weeks that summer I had been watching and wondering at the progressive rotundity of Beauty; it began to worry me finally, for usually it is not very easy to tell on a red deer hind that she is heavy in calf. Yet Beauty grew enormous with a barrel-like girth, but she exhibited no signs of distress, indeed, the reverse, for she was completely red in her new summer coat. I had even remarked to a friend that I had never seen a hind so obviously swollen in pregnancy before, and, facetiously, had remarked 'Perhaps she is carrying twins'.

On the Wednesday of the birth it was raining cats and dogs and, disappointed, I had to cancel my hopes of going to the hill. Back and fore to my windows, to watch if the day faired up, I noted about midday through the heavy rain, that Beauty was lying alone in the deer enclosure and that she was lunging out, now and again, with a hind leg. 'The calf is coming at last,' I muttered to myself. Rain and approaching lunch-time forgotten I went out to keep a watching brief. At 1.30pm Beauty effortlessly delivered a black and slimy stag calf, and with the spring balance which I had ready, I weighed it. At sixteen pounds it was a fair weight but lighter than her hind calf of the previous year. I sighed in relief, glad that it was now over with hind and calf both fine. A midwife, especially an apprentice midwife, is perpetually under strain.

At precisely 1.40pm, while I was watching Beauty busily licking at and cleaning her calf, while she relished too her eating of the birthing 'envelope', my eyes almost popped out of my head as I saw another black-looking leg begin to emerge from Beauty. Do you know, my eyes instinctively slid to the calf which the hind was licking to check if it had all four legs. Disbelieving, doubting the evidence of my own eyes, I watched, gaping foolishly, as the incredible sequence of events unfolded. By 1.50pm her second calf had been born, a breech birth this time, the hind legs and haunches coming into the world first. Another ugly, black and slimy object lay at Beauty's hooves, and, to my dismay, appeared absolutely lifeless. A twin, and stillborn, already I felt bitter about it. Beauty had no such thoughts; transferring her attention from the now struggling, licked-clean stag calf to the lifeless last-born twin, her vigorously licking tongue and envelope-tearing teeth miraculously endowed the motionless calf with life. I raced into the house, dripping wet, and shouted to Margaret 'We've got twins; don't bother about my lunch!' With that I grabbed a fresh film from my cupboard and dashed out again. I weighed the small, black and ugly second calf already partially licked clean. It was a hind calf and she weighed only twelve pounds. This meant that the hind had been carrying, and nourishing, a combined weight of twenty-eight pounds. No wonder she had been so obscenely swollen.

The first-born stag calf was weaving his way staggeringly to the hind's udder within thirty minutes of his birth; his twin sister, very obviously the weaker of the twins, took rather longer relative to her birth time. I cursed the bad light and rain which was seeping through clothes and glistening, in droplets, on my camera. Photos or not, I just could not leave the scene lest this was a dream, a fantasy, a mirage created by my own lifelong conviction that Highland hinds could, very occasionally, produce twins. Why too, should I have been privileged to witness this birth, this amazing achievement of Beauty's? Little wonder that I was so incredulous!

At 3pm, satisfied that I had witnessed both calves successfully have their vital first feed of colostrum-rich milk, I finally acceded to Margaret's plaintive pleadings to come in, get dry clothes on and have a late 'lunch'. We wondered if we would have to hand-rear one of the twins; it would be better, as an achievement which could be duplicated on the hill, if we left it to the hind alone.

In the event we had no hand-feeding to do whatever; Beauty reared both calves unaided and proved her worth again by doing this very well. I had a remarkable hind indeed and she continued to justify this assessment as she grew older although she has never again achieved that particular peak of having twins.

Incidentally, and quite coincidentally, while I was away to the hill at first light (around 5am) on 17 July, a few days after Beauty's twin birth, I encountered a single good-looking hind, not another red deer anywhere within view, followed by twin deer calves, dappled beauties of, I estimated, about one month old. I enjoyed watching them, on a vast, wide, sweep of heather, for some considerable time; I was sure, am sure, that these also were twins, but then, I had not seen them being born.

I used a lot of film on my hind and her twins in the rest of that year. From the day of their birth the twins were looked after assiduously by

Beauty who had abundant milk, as the condition of her twins testified. From birth also they were well nigh inseparable, lying up together, feeding from the hind together, later on grazing together. They lost their infant dappled coat together and grew their first winter coat together. This companionship lasted until, as a two-year-old and with antlers of six points, the young staggie decided he was independent of the all-year-round female company. He began to lie apart from the hind group, to act, in fact, like the deer on the hill. His twin sister grew to be a handsome hind and with the same nice nature as Beauty herself, very tame, and a favourite with our numerous visitors each summer season.

Not all of my recollections of my tame deer at Torridon are happy ones. Once I lost a dappled calf only a week after it was born. Once too I lost a favourite hind, a small-framed, very nice-natured hind, who died while trying to give birth to her calf, even though I had my friend, Roy Peacock, a former vet, helping her. We did actually manage to deliver the calf but it was born dead and the hind, uncomplaining to the last, died later.

The saddest and most bitter of my recollections of my tame deer relates to a tame stag of whom I had had great hopes; not only was he good-natured he was also a beautiful stag who had a royal head at three years of age. My royal's mother was Beauty, who had been the mother of the twins. He was born in June 1985 and weighed twenty-and-a-half pounds. His points, on this twelve-point, three-year-old, head were admittedly short in length, but there were the six points on each antler, brow point, bez point, trez point, and three points (in the shape of a cup or crown) on top, which go to make up the conformation of the royal head. At four years old and still only a young stag, he had a magnificently impressive royal head, balanced and with long points. I had great hopes for him; he would undoubtedly have a most beautiful and memorable head when he had grown to maturity. I should have known better. Unfortunately a gate was left open and my royal found it before I did and went out. He was never allowed to reach maturity, a selfish 'stalker', disregarding all the ethics of deer-stalking, who was 'guiding' (*sic*) a rather inexperienced 'rifle' saw to that.

Happily I have lots of more pleasant memories to recall; the sparkle in a child's eye as the soft moist nose of a hind sought food from its hand; the unbelieving admiration which the sight of a young, dappled deer calf brought into everyone's eyes. The wonder in children's eyes as I showed them how every individual hair in the coat of a red deer was crinkled to aid in insulation; the delight when I let them watch Beauty perform her latest 'trick', that of opening the lever-type handle of the deer larder door because the deer nuts were kept in there. As Beauty grew older food became her obsession to the extent that to leave any door open was to have Beauty in the house, questing hungrily. Fruit, packets of sweeties, a loaf of bread, a container of flour, (which burst open as she pulled it to the floor and shrouded the scullery in white) were all sampled. Bus parties began to take slices of toast from their hotel breakfast to lure her into their bus and she was very willing to be so lured.

Even when she was hungry Beauty was always gentle, so that she could be trusted with everyone, children, their mothers, disabled visitors in wheelchairs, excepting, I must confess, dogs. I usually requested visitors to leave their dogs at their cars for this reason; Beauty was protectively maternal, and dogs looked too similar to foxes in her book. Many a dog went fleeing down the drive, or if small enough, was picked up to cower in its owner's arms, away from the plunging front legs of Beauty. Many a visitor, juvenile or pensioner, with all grades inbetween, from the most industrialised areas of urban Britain, saw and touched a live red deer and, I am convinced, loved doing so. They were told just why we *had* to control and manage such attractive animals; why our policy was not to have deer-stalking practised for sporting purposes; why we could not just allow our deer on Torridon to breed unchecked and multiply since this would lead to overcrowding, damage to the habitat and consequent suffering to the deer.

Red deer were an integral part of the Torridon scene; magnificent, indigenous and adaptable wild animals. Most of our visitors left Torridon with a better understanding of them and their survival, annually, of the bleak Highland winter, quite unsubsidised, and largely without any supplementary feeding in the harshness of those winters. I am convinced that both the deer and their visitors were the better for this understanding.

10. Red Deer on the Hill

Deer, mainly red deer, have been my joy, my hobby, my obsession for practically all of my life. I have stalked deer with a rifle, at first as a youthful challenge, then later, with better understanding, because I believe that stalking with a rifle is the best, most selective and most humane way to accomplish the very necessary management of our wild deer populations. I have stalked them with a camera, even in the 1950s, before the development of the single lens reflex camera and the numerous telephoto lenses so readily available nowadays, (and wasted a lot of effort and film in so doing) for our Highland red deer are truly splendid animals, living in splendid surroundings and present a perpetual challenge to the photographer. I have stalked them with only telescope, walking stick and notebook, at all seasons, simply to enjoy watching and recording their way of life. It was through this watching, this constant presence among the deer, at all seasons, that I became the first one to identity that call peculiar to red deer hinds at their calving season which I came to christen 'the calving bellow', whereas, prior to this identification, this call was attributed to the voice of a young stag, learning to roar – in June!

Perhaps it is appropriate to begin this sharing of my days on the hill with deer in the month of June, for this is a magical month in the Highlands, with new life being born all around one, in trees, in wildflowers, in birds and animals.

One hot June day at about 6 am I slanted slowly up Liathach, pestered by sweat, midges which got stuck in this, and clegs which seemed to relish mixing this sweat with the blood they were trying to siphon from me. Deer, too, were suffering. Red deer, like the golden eagle, exhibit much more discomfort and distress in hot weather than on the coldest, frostiest day in winter. Those deer I saw were standing in groups, or in lines, high up on the skylined ridges, getting what comfort they could from the cooling breezes wafting there. Now and then one of such an assembly would throw its head in the air, long nose pointing skyward as if in supplication, tossing this back and fore wildly as a prelude to dashing up hill to an even higher ridge. The contagion of the solitary hind racing away would spread quickly until that entire group was also racing up to another higher ridge, identical, it seemed, to the one they had just deserted. At times a lone hind, usually an older, more experienced one, would be barely detectable, deep in the shadowed gloom of the north face of a shelving outcrop, shielded thus from the sun. One such hind, a big, dark-coated beast, liberally bedaubed with dried-on peat from a recent wallow, was quite unwilling to be shifted out of the shady gloom of her overhanging outcrop. There was a look almost of ecstasy in her appearance as she savoured to the full her deep, black shade and the coolness of the slight breeze reaching her. Nor was she going to leave; bedamned to that mad human, passing so close, walking out in that broiling heat, when all else was seeking shade. She watched me go by without stirring a muscle and I left her enjoying the cool seclusion of her rock shelter. Another solution to such a hot day in June is to lie sprawled out on any expanse of hardened snowdrift, left from the winter in a north-facing coire, while, of course, the wet, semi-liquid peat of a deep, shadowed, peaty wallow is yet another comforting stratagem. Wet peat when it dries on the coat into a hard, if brittle, skin, protects the more tender parts of the deer to some extent from the various flies and biting insects which pester them in hot weather. Many of the deer on that hot June day were combining the cooling ridges and the comfort of liquid wallows, for their coats were black-looking, with stiff, spiky bunches of hair standing out with dried peat. The peaty pools I passed were clouded from the stirring-up of recent use by the deer, while 'rafts' of loose deer hair were floating on the surface and fringing their hoof-scarred edges. June is a time of year when the deer are loose in the coat, their old thick winter coat, its usefulness over, is coming out, with the thinner sleek red coat of summer growing in to replace this.

Quite a few of the deer were obviously suffering badly from the unwelcome attentions of the nasal bot-fly. This is a type of bot-fly, (*Cephonimyia auribarbis*) which plagues the deer throughout June. About the size of a bluebottle fly it selects a deer and homes in on her, or his, head and attempts to deposit tiny, live 'grubs' on the inside of the nostrils. Each individual tiny grub is furnished with a pair of 'hooks' with which it attaches itself to the sensitive insides of the nasal passages and, infinitely slowly, they make their way up the nasal passages, undoubtedly causing the deer intense irritation, until, eventually, they reach a 'pouch' at the base of the nose and throat. There they 'hook' on firmly and there they live, and thrive, until the *following March* feeding on the mucous there. I have counted as few as one, and as many as seventy-three in a deer's head when checking for this nauseating parasite's incidence in deer, often in poor condition, which I had culled that season.

By the time they are sneezed and coughed out by the host deer, in the following March, the tiny 'grub' which intially became deposited by the bot-fly, is of flattish, segmented section, approximately a quarter of an inch wide and one-and-a-quarter inches long, of yellow-white colour, repulsive in its looks as in its habits. Having been sneezed out by the deer, after its long period of living inside its head, it pupates in the soil. You will realise why I question the purpose of such a fly.

The deer attacked by this fly are obviously terrified by its attentions and individual animals will stampede wildly through the hill to try to evade attack. On that June day I watched hinds under attack toss their heads wildly in the air, shake their heads until it seemed their ears would fly off, push head and nostils into damp spaghnum moss, and 'paw' at a downbent nose, almost savagely, with a dabbing front hoof. Sometimes they would blow violently and very audibly, in the clear hill air, so that one heard the snort at two or three hundred yards distance. At others, a hindleg, outstretched, was used to rub the attacked muzzle frenziedly, up and down, in obvious attempt to dislodge that which was paining them. Last resort was the plunging, bucking, head-tossing run through the hill, possibly the rush of air so engendered soothing the nostrils somewhat. Next time you are bothered by midges in the Highlands, spare a thought for the deer.

In another year I went to the hill early, for the June hills have always fascinated me. The dawning had been bright, perhaps too bright and clear, for by 6 am the weather changed, the day became overcast with a steady north-west wind blowing so that I expected the onslaught of heavy rain at any moment. Blessedly, this just did not transpire. Coming over a ridge, with the wind blowing behind me, I suddenly, and incredulously, saw a single hind, in bleached, faded, yellow-white, old winter hair, lying, curled up, nose tucked well into flank, with only the pointed tips of her ears showing over the line of her spine, and she was only ten yards away. My initial reaction was that she was dead; then, seeing the flanks gently rising and falling, I thought 'she's resting, exhausted, after a hard birth of her calf.' I stood, stock still, was I dreaming; had I really seen her flanks stirring as she breathed? Then I saw one ear-tip twitch, she was alive. I peered hard at what I could see of her rump – no, there was no sign of a half-born calf there. I stole a foot or two nearer and got a better look at her head, tucked closed into her flank. Her eye was closed; she was asleep! Slowly and infinitely cautiously I began to work the straps off my rucsack from off my shoulders, to get my camera. An ever so faint whisper of sound as I finally eased it off and both of the hind's ears twitched. I took out my camera, praying, and watched the ears relax again. A jacket sleeve brushed the nylon fabric of my rucsack; the sleeping hind's head jerked up and around and I found myself looking straight into a pair of amber eyes. Astonished, incredulous as she must have been, her reaction was incredibly fast; she was on her feet and running before I could get a second photo, at every bound a querulous, disbelieving, bark was jerked out of her as she vanished beyond the next ridge.

On this June day too I heard, more than once, what I had come to accept as symptomatic of the red deer calving time, the calving bellow from a hind. This, as I have already written, has often been attributed, in the past, to the rather juvenile 'roaring' of a young stag learning to use his voice, for future needs at the rut. It has none of that intimidating quality of the aggressive, abrasive, throat-rasping roar of the mature stag yet this hind calving bellow could be and indeed often was, likened to the softer bellow of a young stag. Perhaps I am being over imaginative again but to me it had different qualities, nuances, if you like, according to how this June bellowing was used. I could (or thought I could) detect a sense of longing in it, an unrest, when it was used prior to the birth, as if forced to voice this by the 'kicking' of the unborn calf in her belly; I am certain too, that I could detect a most definite, vaunting, self-satisfaction in it, as when the hind stood, straddle-legged, over the feebly-stirring, shapeless, black-looking object which was her new-born calf. I could definitely detect distress, even mourning, in the bellow, when a hind stood over a stillborn calf, or over the body of a calf which had been killed in her absence, by fox or eagle. Once, too, I saved the life of a deer calf, immured in a deep peat cavern, because her mother stood, bellowing thus in obvious distress, over the small hole through which her calf had dropped into an underground peaty chamber.

Later, that June day, on Liathach, I heard, and witnessed, from a distance with my stalking glass, the calving bellow from a young hind as she strove to give birth to her first calf. Quite a long struggle she had, too, reminiscent to me of the first deer calf birth I had ever witnessed for her labour lasted from 11.20 am until 4.15 pm. When I spotted her first she was pacing about restlessly, and I could see a yellow-white hoof and around six inches of shiny, black leg protruding from her rear. Now and then she ceased pacing, and strained hard. Through my spying glass I could just make out a second yellow-white blob of hoof appear, only to disappear as the hind relaxed. She bellowed at intervals in a low-pitched uneasy note. At other times she lay down but not for long. I hoped that it was not going to be a case of a stuck calf for I knew that she would never allow me to get near, to aid her. To my relief (and I am sure to her's) the calf slid, slippery and wet, from her, at 4.15 pm. Half an hour later, as her calf had its very first colostrum-rich milk feed, the mother, unease and travail forgotten, bellowed again, long and low, expressing, I felt sure absolute relief and self-satisfaction.

These then are some of the feelings a red deer hind may give vent to, by using the calving bellow, in her calving season. You may feel I have been over-imaginative in thinking thus, but I have enjoyed many years on the hill among the hinds in their calving season, listening and watching and recording, and I am absolutely sure that a hind can express her feelings of the moment by this bellow and the very different nuances which she can impart to this. Listen for this bellow if you are on the hill in June.

The ear marking of recently-born red deer calves, in their birth month of June, was practised so as to learn something more of the range of hill over which hinds and stags would roam during their lives and also something more of their lifespan and their weights at birth. I found this a fascinating study and gained a lot of knowledge, as compared to guesswork, by this practice. It is now a known fact, mainly because of this ear marking, that red deer hinds are very tied to their home range; most red deer hinds live and die around the same area of hill in which they were born. The stags are a different matter; they have no ties, no paternal instincts, no responsibilities and so, footloose and fancy free, they roam much more widely. Individual stags have been recovered on an area of hill far distant from where they had been born, as far as eighteen to twenty miles away.

By the time June ends most of the red deer calves will have been born and quite a number of those born early in June will be showing now among the hinds adding attraction and new life to the hill. Occasionally then, too, you will see a dappled calf wandering through the hill alone, or perhaps a couple of these together. Like children, deer calves vary in temperament and, as in children also, this can be dangerous at times. The mother hind is a courageous, selfless protector to her young calf and will thwart eagle and fox should they try to take a calf while she is with it. An impatient calf left lying safely immobile in some heathery corner, or shielding peat hag or large rock, is relatively safe, even from the ultra-sharp eyes of the eagle who, at this time, is also feeding young. Wandering through the hill, minus the hind's protection, renders the calf fair game. I was told reliably of the following occurence on Rum, where they still ear tag deer calves for research purposes. A deer calf had been caught, weighed, sexed and ear tagged. The researcher left the calf and walked some distance away then waited at a vantage point within binocular range to watch for the return of the hind. Mist came down and the wait became a cold and fruitless one because little could be seen of the surrounding hill. The 'watcher' stuck it out patiently although chilled to the bone; the mist, as suddenly as it came down, cleared and where the marked calf had been left there was now a sea eagle, having a good meal from the dead calf. I have personally witnessed deer calf brought in as prey (most often carried in as two separate halves, the eagle dividing them as neatly as any butcher at a mid-spine joint) quite commonly while watching eagle eyries. The fox too commonly shows deer calf remains at its den. I am quite sure that some of the remains seen will have been of calves found dead and that most, if not all, of the others will have been killed by fox or by eagle because they are alone.

Life is hard and unforgiving on the hill and the unwise, or the too enterprising, seldom live to make old bones; the weeding out of the unwise, unheeding or just unlucky, of any species, begins from youth. Red deer hinds are almost invincible in their maternal courage at this time of year and many deer calves must owe their survival to the maternal protection. I have known of foxes routed by hoof-lashing hinds; have watched an eagle being forced to take wing from its vantage point on a green knoll; have heard of an eagle, slow to take wing from a calf it had just killed being literally trampled into a feathered pulp by the hinds which came running to the rescue on hearing the far-carrying panic squeal of the dying calf. Shepherds' collies have been chased while on the hill by a maternally-motivated hind; my tame hinds at Torridon regularly chased visitors' dogs when they had new-born calves. One of my hinds I finally nicknamed the Shrew because she was so vixenish and aggressive when she had a calf, to the extent of dealing me many a shrewd blow with a hard hoof while I bent down to weigh her new calf, once, in fact, ripping my jacket. Quite innocent intruders sometimes incur a mother hind's protective wrath; she has no antlers but her plunging forelegs, hard-hooved, are effective weapons. A couple of nearly full-grown otter cubs, while journeying across a deer park, via a rushes-grown ditch, came, unwittingly, too close to a recently-born deer calf. The hind killed them both, fracturing their skulls. A hill shepherd, a reliable chap, once saw a hind with a new calf rising up on her hindlegs and plunging downwards with both forelegs at something too low in the heather for him to identify at a distance. He moved in, after the hind had gone with her calf and found an adder with a broken back. I have had the experience, once only in many years of looking for and ear tagging deer calves of a hind who refused to go away even as I moved in to catch her calf. She moved away perhaps ten feet and stood there watching as I sexed, weighed and ear tagged her calf and was back licking her calf before I'd gone twenty yards away. She showed every symptom of anxiety and fear, rolling eye, gritting of teeth together, very audibly, tossing of head, yet run away she would not. That, to me, was the ultimate test of a wild red deer hind, this standing of her ground to the most feared species of all, a human being.

Calving time over, there is still plenty of interest on the hill. A group of dappled calves, as full of youthful *joie de vivre* as any spring lambs, playing the identical game that these same lambs do, king of the castle, racing around an eroded mound in a dry peat bog, one of their number acting king until another jumps up and displaces it. Or in early morning, an elderly hind, acting, it appears, as babysitter, for around her recumbent but alert form, are scattered no less than five dappled calves. Quintuplets? Hardly likely! Or a hind, all on her own, who completely thwarted my planned approach to a favourite coire. Fate, in the quite unexpected guise of a single, gaunt, red hind, etched stark against the brightness of the morning sky, on the distant ridge. Tiny, a dark silhouette, skylined, yet absolutely omniscient in her absolute command of all the hill below her. If disturbed, turning in alarm from her fortuitous vantage point, she would have disturbed the entire coire I wished to visit. By her very presence I was forced into a long wearisome detour.

The habit of our Highland red deer, both stags and hinds, in chewing cast antlers and the bones of any dead animal found on the hill, is well authenticated. The reason for this habit is, logically enough, attributed to a craving for trace elements and minerals, such as calcium and phosphates, not readily available in their largely acid hill grazing. One can readily understand why a calf-bearing hind should feel this craving since she has to 'make' bone in her production of her calf. The reason is not quite so clear in the case of the stag until one realises that each stag casts, and has to regrow, his antlers annually. These antlers are of bone and not of horn, as in the horns of sheep and cattle, and therefore stags also have to produce bone annually.

Every autumn the annual rutting season equalled the summer's calving season in its interest and spectacle; a rather different kind of interest you may feel, certainly much more boastful and vaunting. Instead of hinds bearing their new calves, and, in doing so, venting their feelings in their calf-bellowing, one had an influx of stags into the Torridon hinds, each and every one trying to assert, by boastful roaring, his superior claims to these hinds. Mind you, these males, noisy, arrogant and blustering as they were, were fulfilling a necessary function (just as in the human race); but for this noise and braggadocio, this strutting and advertising of their virility, there would be no calves for the hinds to care for in the following June.

This mating urge however is the limit of the red stag's involvement; his very offspring are forever anonymous to him. However, each rut was memorable by the very presence of the stags, their roaring progress through the hill in search of hinds, their giving to these normally silent hills a fitting voice which seemed to emphasise their wildness and isolation. Heard in an October gloaming, as one passed, unseen, close by a roaring stag, the sound was elemental and primaeval and indeed, could be very intimidating, amid the darkening hills. One almost felt like tip-toeing past, unwilling to draw the attention of the roaring stag.

With the onset of the rut the entire demeanour of the mature stag (who has spent all summer in reasonable amity with his companion stags) alters dramatically. His very gait alters, from a free striding grace to a stiff, stilted strutting, implicit with menace. His expression changes perceptibly from that of near bovine placidity to that of very tangible, sullen, belicosity. He appears chest-heavy, with enlarged, shaggy, muscular neck, looking even thicker with its hairs stiffened and spiky with his wallowing in peat. His former sleek, summer-red coat is now filthy with this same peat wallowing and his pale belly hair has a black middle line, of curled hair, soaked with his copious spraying of his own urine. Because of this he stinks appallingly, a rank and powerful scent. He is a beast transformed and not only in appearance, his entire personality has changed also; he becomes a beast possessed, obsessed, violent. Might, his own might, in achieving the possession and utter domination of hinds, is his sole purpose in life. It is not really an enviable state; for about three weeks he exists thus, living on his nerves and urges, these jangling nerves preventing him from either eating properly or resting. Each individual stag, in the competition of the numerous like-minded stags on the open terrain of our hills, will lose around two stones in weight. His stomach will be empty at the end of his rut but for a little evil looking black 'fluid'; every vestige of body fat achieved in his summer grazing (and this can be very considerable) is gone. He therefore begins his winter in parlous state, in marked contrast to the good condition of the hinds which he has harassed for all of the rutting season. This is why the venison of a winter stag, in our Highlands, is not worth cooking, far less eating, for in the privations of winter, coming so soon after his hardly idyllic rutting excesses, the stag seldom regains condition.

Fights between rival stags (skirmishes is perhaps a better term) were frequent, even three cornered fights where two contesting stags were charged, broadside, by a third stag wrought up to a pitch where he just had to take action. Seldom did these skirmishes last more than seconds. For a fight between rival stags to last long enough to be able to get within close quarters to watch it, both combatants had to be matched in weight as carefully as for any boxing championship, regardless of what kind of antlers they possessed. The deciding factor in this relatively rare kind of rutting battle between two equally matched stags was, after a long, dour, bitterly contested struggle, possibly a little slice of luck, when a misplaced hoof slipped or slid, or that all important ingredient, whichever stag had more fire in his belly. I was only privileged enough to be able to watch one such fight on Culachy in nineteen years on the hill there, and I too was absolutely drained after watching that Titanic struggle. Similarly at Torridon, in twenty-one years, I was only to witness one really long-lasting, memorable fight of this class, and to photograph this fight.

Like athletes, stags go into training, albeit instinctively rather than consciously, for their annual competition which is the rutting season. The mature stags will have their 'new' antlers grown and hardened off well before the rut. Armed with their new antlers there must be an instinctive urge to get to know these new antlers and this obsession grows stronger as the rut grows nearer. A stag, relearning the use of his hard antlers (which he has had to renounce for about three months) prefers to sharpen up, as it were, on inanimate objects, something whippy, elastic, resilient, which yields before his onslaught but which rebounds back into the fray as soon as the stag slackens onslaught. Favourite 'sparring partners' are resilient saplings or low-hanging whippy branches of trees. Just as in combat, he spars about simulating the preliminaries of getting antlers clear and foreheads in contact. This achieved, the stag pushes forward, meeting the token resistance of low hanging branch or whippy sapling. The action grows more and more furious as he becomes really aroused until it is a hoof-plunging, berserk, full scale rampage, with retreat, parry and counter parry, until the carefully chosen victim of his power play is reduced to pitiable ruin, broken, splintered, barkless, while the ground is a porridge of ploughed up hoof marks. The victorious 'stag' may indulge in a mad, bucking and prancing, curving run, back humped, head tossing, away from his erstwhile 'opponent' (perhaps preparing for defeat as well as victory) before settling down, a placid herbivore once more, as distinct from a whirling Dervish, to graze peacefully. Now and again a stag becomes adorned with a chaplet of whatever he has been attacking, sometimes (if he has been unwise in his selection) of a tangle of flexible wire.

Incidentally, fighting stags, when they get to the goring stage of a contest (at which stage the beaten stag must, weakened badly, retreat fast or risk being, perhaps fatally, stabbed) use their brow points, those lethal points directly above the forehead. Using these points they can see (looking along the line of them, as it were), precisely where to direct them. They do not use their terminal points! In order to use these they would have to have their heads so bent under, nose almost between forelegs, that they would be completely blind to the direction of a thrust. The hoary myth, that the switch is a killer stag because of his long, apparently lethal, spikes is therefore, while being perfectly understandable, also a very definite myth.

On that October day when I watched the well-matched contest between two stags on Liathach I had been watching a seven-point stag, who, roaring vaingloriously, was holding a few hinds. I suddenly realised that I was hearing that staccato click and clatter, almost metallic which tells of a couple of stags fighting. The noise, far-carrying in the clear air, was high above me, out of sight behind the series of ascending rocky ridges. I slid quickly out of sight of the seven-pointer, rejecting him utterly, and then began a lung-bursting, leg-tiring race to get up and over the intervening ridges. I may add that I could run uphill in those days, for this happened in 1978. Incredibly, after years of frustratingly brief skirmishes, I eventually got within sight while the combatants were yet consumed and concentrated entirely with their fighting. Nor could either afford to slacken concentration for even a split second, for the contest was Homeric, antlers locked, foreheads hard in contact, eyes bulging, pink tongues lolling half out of mouths. Whirling and wheeling they were, each trying to win the temporary advantage of the uphill stance, thereby pushing downhill. The whirling of one stag in order to gain the upper ground, was almost immediately countered by a twirling, twisting countermove of the other. In the blur of the action it was impossible to count the points on either head. Scattered around, as interested onlookers or indifferently chewing a restful cud, were the hinds for which this truly Titanic struggle was taking place. Knowing the fighters were absolutely engrossed in their battle for hinds I ran straight for them, my telephoto lens scarcely necessary. The hinds, both those interested and those indifferent, ran, of course; what did it matter to them anyway, there would be another wildly roaring stag just around the corner. Still incredulous, I tried to control my limbs, shaking as much from excitement as from exertion, and crept nearer to begin my own type of 'clicking'.

The fight raged on and on, the usual pattern of a strength-sapping pushing match conducted on terrain which was so steep and rocky as to make it difficult to walk never mind fight. So nimble, so surefooted, were those two stags that they seemed feline rather than cervine. Long seconds intervened when all movement was sus-

pended, when each stag was graven in ultimate, supremely con-
tracted endeavour, in a state of suspended animation, every rigid
muscle on the taut straining haunches bulging. Then, in an explosive
burst, one or other of the stags would draw on some miraculous inner
reserve and with a tremendous expenditure of awesome power would
propel his rival backwards, a splatter of mosses, grasses and
fragments of peat erupting from deep-gouging hooves. Now and
again a truce seemed declared by mutual agreement. Both stags
would extricate their antlers, withdraw backwards and, keeping
antlers threateningly lowered, retreat a few yards, and then charge
each other again, in an incredibly fast, short savage rush. There was
by this stage no delicate sparring of antlers in order to get foreheads
in contact. These two stags now rushed at each other with insenate
venom and antler-splintering violence. It was abundantly clear that
the eventual loser must rush away fast, for dear life, or suffer the
brow points of his inflamed rival in his body, somewhere.

It was not to end in victory for either stag; so absolutely evenly
matched were the gladiators (all the best fights are between two such
opponents) that both stags tired simultaneously. They did not accept
this stalemate at all gracefully, it was an ungracious, blustering,
roaring stalemate in which each stag, after separating, stalked stiffly
away from the other, rump to rump, then halted, thirty yards
separating them, and one roared vituperation at the ridge of Liathach
to the South, while the other roundly abused Beinn Dearg to the
North. Though I was in full view I was ignored as if I was magically
invisible. A final outburst of roaring abuse and then each stalked
away, in the direction each had chosen to roar at. Quiet reigned once
more except for the far away soothing symphony of the high waterfall
falling out of the verdant green coire below Mullach an Rathain. I
had finished all my film and I too stalked away homewards.

Despite the apparent ferocity, fatalities, even on ground so steep
and rock-shattered as is Torridon, are few, and, I believe, some of
those fatalities are fortuitous, though I do not for one moment
believe a stag who becomes victor by one of these fortuitous accidents
is in any way either regretful or remorseful. Early on in my years at
Torridon I found a dead stag in November one year, below a sheer
outcrop on Liathach. One antler was thrust so deep and hard into the
peat and heather that it took very considerable effort to pull it out. I
believe that in the usual shoving match of two competing stags they
had been so near to the sheer rock cliff that the one I had found had,
literally and inadvertently, been pushed over. My younger son,
Michael, while stalking on Torridon narrowly missed seeing such an
outcome on Beinn Dearg. He had begun a stalk, which entailed a

time-consuming approach, on a stag high on Beinn Dearg, a stag who was holdings hinds not very far from a second stag, also with some hinds. By the time he got up there was only to be seen the stag he was stalking, who now had all the hinds. Successful in his stalk he was dragging the carcase down, threading his way through rocks and screes, when he came upon the other stag, dead, and so shattered of bone and body that it was obvious what had happened. He, too, had fallen, or been pushed, over one of the numerous sheer rock outcrops, in a furious onslaught in which either or indeed both stags, could have gone birling out into space.

And so to winter which in the Highlands can exhibit breathtaking beauty cheek by jowl with heart-searching hardship. On a bright January morning, intensely frosty, and with a depth of powdery, dry snow of around seven inches after an overnight fall, I ploughed a wavering way through the virgin white landscape while, in the wind which is ever present on the tops, a ragged banner of snowy spindrift blew perpetually. Group after group of hinds I watched, dark against the pure white. I was wearing a suit of white so as to render me to some extent less obvious. Most of the deer were still foraging, working a slow way to a sunny slope where they would lie up for the day. Fascinated as always, I watched that lovely, ever so fluid motion of one foreleg with which deer clear snow from potential grazing. One or two deer were already lying, snugly enough it seemed, on a snowy couch, chewing a slow cud. I was happy that I had finished my culling programme and very content to use my camera instead. I moved cautiously on, on feet which grew colder and colder, halting whenever I liked the posing of my models, the deer, in the sunlit snow, my gloveless fingers sticking tackily to my camera. With my feet unbearably chilled, I left leaving strings of deer still working out into the sunlight. Even in snow they were hill-wise and were keeping to the harder ridges. Now and then a calf floundered through a powdery, deep drift unwilling, as all youngsters, to follow their elders. I knew the thin dark lines of deer would wend their way in, in the frosty evening to come as night drew in. I sympathised knowing that my way lay homewards along the snow-covered road, still without any tracks, except for my own and those of the deer, to a warm fireside.

Before finishing the chapter I would like to write a little about the often controversial subject of stalking.

It is my personal belief that stalking is the best management tool that we have, to ensure that our native wild red deer do not increase in numbers to the extent that they do themselves disservice in eating out, or otherwise damaging, their habitat. The Deer Commission, charged in Scotland with, significantly, the dual role of the conservation and control of red deer in Scotland, emphasises the point. 'The red deer is our largest native land mammal and when the last wolf was killed the last natural predator died. It is therefore necessary for estates to fulfil this essential role and to keep numbers under control by killing, by stalking, such a proportion each year as is necessary to keep the deer population in balance with the grazing available.'

In stalking red deer, for control and management purposes, to their benefit as well as our benefit, we are filling the former role of the wolf in weeding out those deer, of either sex and of any age, which are, for one reason or another, unlikely to survive the winter, with all its hardships. In many cases these will be old deer which, after years of progressive wear on their teeth, can no longer graze efficiently enough to sustain life. Deer do not get sets of false teeth. There will be deer which have sustained injuries, or are not thriving for any other reason; there will be below par youngsters who may have been born in the preceding summer. Deer calves which weigh less than fourteen to fifteen pounds at birth seldom have a chance to reach maturity, or to survive the hurdle of their first winter. By pruning out these deer one is ensuring that they do not die the terribly slow wasting-away process, towards the end of winter, which is called 'natural mortality'. This so-called 'natural mortality' is neither more nor less than a convenient euphemism for starvation, for weak deer cannot compete for grazing with their healthy companions. In a species with a very strong pecking order the best grazing goes to the fit, strong deer. This does not matter in summer and autumn with an abundance of grazing; the crunch for weak deer comes in winter and early spring when grazing is minimal.

Ideally, in a well-managed deer forest, there should be almost no deaths in early spring, because the weak deer should have been pruned out. These pruned out deer will be saleable, even though thin, and so will provide some economic return; if allowed to die, of slow starvation, they are, very literally, a dead loss. Dead or dying deer, increasingly noted by hill walkers, are not good publicity for our 'management' of our red deer.

Forget about stalking as a sporting pursuit. This is incidental. Should stalking for sport cease completely tomorrow, then stalking for management and the well-being of our deer would have to continue annually. The dearth of grazing in the long winter and early spring period in the Highlands is the critical factor which our native red deer have to face annually. The winter coats of our deer are well adapted to withstand cold, with wiry outer guard-hairs and an inner woolly underlay to retain warmth. There exists, however, a very fine-honed balance between the numbers of deer present in winter and the inevitable paucity of the food which is then available to them. This is when stalking and wise control by us, (enlightened, we hope) humans, comes in and this annual control is vital, for red deer and habitat. The surest way to destroy any species is to destroy, unwittingly or not, its habitat. I repeat, stalking is vital.

Deer stalking should, as conducted on well-managed estates, entail the absolute minimim risk of suffering to the quarry. Skilled professional stalkers either take the shot or guide the guest. The proficiency of a guest is ascertained by a session with his rifle on the target before he goes to the hill. Rifles nowadays are highly efficient and accurate tools, equipped with telescopic sights for precision shooting, vastly more efficient than in the early days of deerstalking, when effective accuracy range was about eighty yards and black powder obscured your vision for a moment or two after the shot. Wounded deer, which require a second shot, should be reduced to a minimum nowadays and wounded deer, which escape to die later, to insignificance. No one who is not a natural good shot should even contemplate the role of the professional stalker. Only a single incident with a wounded deer should be *more* than sufficient to emphasise this precept. A badly wounded hind, lying down, quite unable to regain her feet, head low, instead of proudly erect, ears beginning to droop, eyes hurtful in unvoiced bewilderment, their lustre slowly fading; neither complaint nor resentment is voiced. This is how wounded deer die. *You* have been responsible for this; this alone should motivate you to *ensure* that you shoot well – *or to give up deerstalking*.

This, then is my avowed attitude to deer stalking. It is not intended as an apologia; one does not apologise for actions which one believes are correct. I hope that it is the well reasoned, responsible attitude of a fairly imaginative individual to a species of wild animal which he esteems and admires. I have *no* pleasure in spilling blood; in every hind season there arose a situation in which I asked myself 'why' I was dealing with dead deer, with, literally, bloody hands. Was there not a preferable way of life? Only reason, human reason, came to my rescue at these annual bad moments.

It was while I was still at Torridon that I received an almost unbelievable invitation from His Royal Highness, the Prince of Wales to stalk at Balmoral. My initial reaction was one of stunned disbelief; however, there had been *no* mistake and across to Balmoral I duly journeyed, from the west of Scotland to the completely different east of Scotland. I stayed with Martin Leslie, who was factor for Balmoral and there I was shown every kindness while Martin also did his best to banish my nervousness. Nevertheless I was still nervous next day since I was now aware that Prince Charles had said that he himself would be my guide and stalker.

When we did meet up that morning (a nice, dry morning), Prince Charles quickly had me at ease. We had two ponies of the very handsome, yet workmanlike, Haflinger strain from Austria, and with them, two young soldiers to handle them. Another young soldier was to accompany us as a ghillie and he carried a walkie-talkie set so that later he could direct the ponymen who also had a walkie-talkie set. The ponies were equipped with the combination-type deer saddle, so called because they could serve as riding saddles as well as deer saddles to carry the deer carcases home. This too was different to the saddle we use in the west, called the Glenstrathfarrar saddle, which is a type designed to carry deer only. One can ride on it but it is exceedingly uncomfortable, even if one tries to ride it side-saddle.

We set off through the Ballochbuie wood, a truly magnificent remnant of the old Caledonian forest with some of its trees aged at least three hundred years. The forest floor was typical of the old open pine forest, heather, blaeberry and mosses, quite distinct from the deadness and sterility of the modern commercial spruce plantations. Two stags, unseen, were roaring, one to either side of our way up to the hill. Once a handsome red-brown hen capercaillie winged, strongly, down into the trees and once we passed the conical, pine-needle heap of the wood ant, a-swarm with frenetic ant activity.

Midway through the wood Prince Charles asked me if I would load my .243 rifle. I enquired whether he preferred a bullet 'up the spout', or all the bullets below the bolt in the magazine. He grinned at me and answered 'What do you normally do?' I replied 'One up the spout'. The Prince said 'Do that, then. You will have a reliable safety catch?' I nodded and loaded up. I then had the further embarrassment of not being allowed to carry my own rifle; Prince Charles was stalker in charge and he insisted in carrying the rifle. I felt rather naked, going stalking and without my rifle on my shoulder. We worked our way slowly and carefully through the wood, Prince Charles spying at each clearing. As the trees gradually thinned out a heather ridge loomed above us. Three hinds were on its skyline and, spotting us, immediately ran down towards the trees – woodland deer habit, whereas, on the hill, without woodland cover to make for, the deer would have made out higher. Out of the wood we cut up to the top of the ridge to have a comprehensive spy from its eminence. The visibility remained good and I managed to find a big mature stag, with a poor type six-point head, lying on a distant rounded, heather ridge, that type of rolling ridge which is difficult, at times impossible, to stalk in to a stag. He had a young staggie lying near to him. The mature stag had no hinds, he was therefore a 'traveller', still looking for hinds and, as such unpredictable, in that he was liable to rise and go onwards, seeking hinds, at any time. I would dearly have loved to go for him despite this but Prince Charles decided to look further afield and one does not argue with your stalker, especially a Royal stalker.

We carried on out into the hill and almost trod on a grouse which flew, meanwhile exhorting us to 'go-back', 'go-back'. On his flight he put up a little staggie which suddenly sprung to his feet from a tiny hollow ahead, where, lying down, he had been quite unseen, and ran out into the ground we had yet to spy at. Unpredictable bad luck, as is always a factor in stalking; we had no way of knowing he lay ahead of us. We gave him a few minutes grace before we carried on to the next ridge ahead. Grouse again, a covey this time, rose ahead of us and curved their way around the ridge; we moved on to that ridge, where we sat for a spy of the wide expanse of ground thus revealed ahead of us. Our spy revealed that there were four mature stags lying in the distance where a wide flat opened out, comfy in their heather beds. On his own, around a mile distant from these stags, a lone stag, another traveller roving in his search for hinds, was coming across our front, slanting slightly towards us. He was still so far away that the only thing I could detect about his antlers was that they were spike-topped (i.e. no cup or fork) and now and then he paused to roar his unrest to the hill.

We lay watching for some time, comfortably ensconced, spying at the various stags and commenting on their respective heads, in the way of stalkers since stalking first came into vogue. Suddenly one of the group of four stags rose to his feet to stare towards the travelling stag. One by one his companions arose also and then, into our view, appeared our young staggie, on a line which would bring him near to the mature travelling stag. So slight an incident on the hill, an unseen straggie, frightened by the alarm of a single grouse, can, like a flung stone on a smooth lochan, generate widespread ripples. Our four stags were not terribly worried but, safety first obviously in their minds, they did not resume their beds in the heather but walked slowly away, picking a mouthful of heather, every so often, as they did so. The travelling stag, ruled by his obsessive urge for hinds, did not follow them but kept on steadfastly in our direction. Our staggie, unpredictable as ever, opted to follow in the traveller's steps. We for our part let both get out of sight and then began moving fast to where we would get across the traveller's route. My stalker judged the interception angle perfectly and at last there was only a low swelling ridge ahead, from the top of which we should see our traveller. Prince Charles progressed ahead of me on his belly, and pressed himself even more into the heather as the ridge top neared us. At the top, while I lay still behind him, he used his binoculars to view the approaching stag, appraising his antlers for what was, to me, a nerve-wracking length of time.

The policy on Balmoral was as mine; bad heads were to be taken as a priority while any head likely to improve was always to be left. I was sure that the head of this stag was never going to improve; spike tops seldom do, after maturity. However I was more than content to abide by the decision of my stalker, although, by now, I was consumed by the sole desire to share a stag with Prince Charles. Mind finally made up, the Prince wriggled slowly back and began taking my .243 out of its cover. 'Rather a borderline case' he said, 'he'll be head on to you but he will be in fair range.' I in turn wriggled forward, to the side of a flat rock, and cautiously appraised the situation. Our stag was halted, around eighty yards away, looking it seemed, straight at me. My dread then was that I, and I alone, would make an infernal mess of it. I have never suffered from that type of nerves called 'stag fever', but I was close to it then. Steadying myself mentally, I placed the cross hairs of my rifle 'scope on the middle of the chest of the stag and carefully I pressed the trigger. I actually heard the comforting sound of the bullet striking, saw the stag whirl around fast, go out of sight, re-emerge into view again, halt, then tumble down with finality. Almost simultaneously the young staggie hove into view and began to circle the fallen stag as if in wonder that he should choose to sleep there. True to type, he lingered long enough to have us cordially detesting the sight of him once more, then, possibly influenced by some occult thought transference, or, more likely, by our waving hats and handkerchiefs under his interested gaze, he made reluctantly away.

The dead stag did have long spike tops, devoid of other points on these tops, but he also possessed good brow, bez and trez points. I was absolutely confident in assuring Prince Charles that this stag had never had, nor would ever have had, the good top points to complement his nice lower points for, by his teeth, he was now twelve to thirteen years old. I sincerely hope that he believed me.

I took out my knife to begin to gralloch our stag. Once again I came under my stalker's veto. Prince Charles grinned that most engaging grin at me once again and produced the knife which we, of the British Deer Society, had presented to him, and said, 'You can make an eye witness report to the Scottish Council of B.D.S. that I

'Beauty', a tame red deer hind, cleans her twins soon after their birth.

An affectionate 'thank you'; the author with 'Beauty' and the twins.

A roaring stag announces the rut.

A stag holding his hinds, relaxed in the late autumn sun.

posite: *One of my saddest memories is of 'my' four-year-old royal which was shot by a selfish stalker.*

A hind and her last year's calf by a boggy pool in October.
A hind making 'the calving bellow' while her hour-old calf lies hidden.

In the late summer stags lie on the high ground awaiting the rut.

Early snow on the tops, as the author spies in typical Torridon stalking country.

My most unexpected visitor was this eagle which landed at our house.

'I heard the tips of her inner wing rasp on the rock'; the hen eagle attacks the author.

The hen eagle with her two-week-old chick.
An eaglet wing-exercising prior to its maiden flight.

An almost fully-fledged eaglet showing its massive talons.

am using my knife'. With that he gralloched the stag as neatly and cleanly as any stalker could have done and, indeed, better than some stalkers. With an air of bliss I hoped not too visible about me, we waited for the called-up ponies to arrive. The stag was loaded, the ponies departed; we lunched, relaxed in the fine weather, enjoying the wide spaces and clean air and blethered about deer and deerstalking.

Lunch ended, Prince Charles was keen to press on further out, toward Lochnagar. Away we went over a wide heathery flat towards a distant and much higher ridge. Lochnagar now loomed ahead, to our left front. A brace of ptarmigan, white-winged, grey-bodied, flew ahead, to curve around a shoulder of the hill. It was about that point that my troubles began; for the first time in many, many years I began to suffer cramp in my thigh muscles and I found myself struggling to keep up with Prince Charles. The fact that I could hear the soldier/ ghillie, the youngest of all of us, drawing laboured, panting breath behind me was no real comfort. I was, at fifty-seven years of age, now at the stage when I really understood, and painfully appreciated, the then enigmatic remark of a red-faced veteran guest, whom I had out stalking in my youth to wit 'At one time these days on the hill, stalking, were all joy and pleasure; now it is all pain and suffering.'

The wind had risen to near gale force; my stalker decided (perhaps he had sensed my travail) to have a spy of the ground ahead. Thankfully I lay beside him, our glasses necessarily steadied on huge rocks in that wind. As we did so a roar from not too far galvanised both of us. A crawl forwards, more rock-littered hill ahead, and we found the roaring stag, among these rocks. He was a mature stag with a very nasty switch head, spiky antlers devoid of points. No border-line case this and in unanimous assent we began the necessary detour. Cramp hit me again like a physical blow, my muscles knotted so that I could not take a full stride. I had to halt and knead and punch at the tight muscles; Prince Charles came back to me, grimaced sympatheti-cally, and gradually the cramp eased. We got going once more, working in to get above our stag. The hill here was a chaotic litter of huge grey rocks with no clear view of more than twenty yards. We left the ghillie behind a large rock and began to slither a way between the rocks. A stag, grazing, showed only the line of his back, just a hundred yards away; for a moment he lifted his grazing head and we saw that he was a mature spike-topped six-pointer. A poor head and we could easily have taken him but, no word spoken, we both agreed that the switch was the stag we wanted. A cautious withdrawal and, safe again, we made another approach. Again, in that horrible jumble of rocks, we saw the stag's body first, head hidden as he grazed. I don't know what alarmed him but suddenly his head came up, he gazed uphill and took off. There was no point whatever in spoiling an enjoyable day by taking a risky shot and that was that. In retrospect I believe, that perfectly naturally, our completely inexperienced soldier/ghillie, suddenly left on his own, in a howling gale, high out on the alien hill, began to wonder at our long absence and peered out over his sheltering rock. Skylined now, the sudden movement caught our stag's eye and he was away.

It was getting late in the day, we were a long way out on the hill, it was time to head home. Our ghillie duly collected we started on the long walk back. Once, on the way back, I again asked to relieve Prince Charles of the weight of my rifle. Again I was turned down. Caution and concealment no longer required we talked all the way home. The ponies had already been radioed to turn for home and were by now well ahead of us. With obvious sincerity Prince Charles spoke of how much he enjoyed Balmoral, the hills, the deer, the peace and quiet, the relative informality. I felt entirely en rapport with him and indeed very much in sympathy with him. We arrived back at base at ten past six and there, at Martin Leslie's house, Prince Charles left me, leaving me with the impression that, in his Norfolk-style tweeds, dressed for a day on the hill, he was happier than on many a more formal occasion. For myself, that day remains unforgettable.

11. The Eagle Years

My lifelong interest in studying the golden eagle, its nesting habits and its range of prey, began in 1957 and continued just as long as I was fit to go to the hill. In my home area, around Fort Augustus on Loch Ness, I knew a shepherd who described to me the eyrie of the golden eagle which he knew; 'it was huge; there must have been a cartload of sticks in it.' This shepherd had also chased, literally chased, a golden eagle, along a glen bottom, with his walking stick, swiping at it as it strove to get airborne. That particular eagle had had such a feed of mutton, from a dead sheep, that it had to get to an elevation, such as the top of a knoll, before it could take wing. Such a knoll it did get to, still a step or two ahead of the shepherd and, throwing itself off the top, got sufficient air below its great wings to enable it to fly.

Such yarns kindled my imagination and I was young, fit and living in eagle country. The eagle, of course, has always been prominent in Highland mythology and tales were numerous, quite sufficient to stir the imagination of anyone and certainly sufficient to kindle mine. My mother, too, was a Munro and the Munro clan badge was a golden eagle. Enough said!

I shall always remember the very first golden eagle I ever saw. It was in 1957, in late April, and she was sitting, incubating her almost hatched eggs, on her remote eyrie in a long, steep-sided glen. Quite safe, one would have imagined, from any human interference. The eyrie was on an estate high above Loch Ness. I had been asked to help in looking for fox dens out on the high ground of this estate. Around midday, walking precariously along a steep rock-studded hillside, it was decided to 'look at' an eagle's eyrie. Unaware of what this really meant I welcomed this suggestion. The sight of that eagle, incubating, in the middle of her wide nest, her head yellow-gold as she turned to peer at our sudden appearance, remains vivid in my mind. So also does the sight of my companion's gun leaping to his shoulder, and the sound of the bang which sent the huge bird, still motionless, still peering at us with piercing dark eyes, set deep below frowning eyebrows each side of a formidable beak, tumbling, lifeless, off her nest. One second, alive, vital, imperious, striving to produce new life; the next, only a falling, wing-flopping, dishevelled bundle of feathers. Devoid of life and dignity by the bang of a twelve-bore shotgun.

Perhaps the worst irony of that particular episode was that I liked and respected my companion for his ingrained knowledge of the hill and the deer. His faults were those of the diehard old school to whom the issue was absolutely black and white – they were employed to look after grouse, deer and to protect the sheep of the estate from predators. The eagle killed grouse, on occasion red deer calves, and on occasion, lambs. They, therefore, killed eagles. That outlook was widespread in his generation. Unfortunately it also entailed perpetual warfare on every bird with a hooked beak; merlin, kestrel, sparrowhawk, peregrine, buzzard, all suffered. Having said that, they were sincere if misguided folk with motives far above those of the egg stealers who also battened on to many of our birds of prey. Influencing the outlook of keepers, too, was the fact that their livelihood depended on their ability to show good numbers of grouse. The majority of their employers wanted to be able to show plenty of grouse on their properties and were, at best, ambivalent in their attitude to birds of prey. Education, wealth, property ownership, did not, and does not, always imply intelligence and understanding. Many employers in that era connived and encouraged persecution of raptors and indeed procured poison so that this could be used, quite illegally, in their persecution. I knew of one owner of a large estate who had been on the hill all day, stalking, and had missed a stag. On the way home, the crestfallen owner and his stalker spotted an eagle, perched, digesting, on a crag. A tiny target compared to the stag he had missed earlier that day – he did not miss the eagle, and returned home a happy man.

There exists a much better outlook nowadays, yet one suspects that there are still diehards who flout both law and logic.

That, then, was my first close to sighting of a eagle's eyrie, under circumstances which I could never have envisaged.

I have written extensively on my years of eagle watching in my earlier books, *Highland Year, Highland Deerforest*, and *The Ways of an Eagle*. Some of my later memories, after my move to Torridon, in 1969, I give here, for I kept up my study of eagles, both at Torridon and in my former haunts of Inverness-shire. Most of my weekends each year, after I moved to Torridon, were spent, during the eaglet rearing season (from late April to late July), in journeying to, and visiting, my study eyries in the Loch Ness area. Some of my experiences in these absorbing and arduous years, when I was well past the flower of my youth, I recount here. Much, of course, was repetitive, some experiences were, shall I say, memorable. Such a one was the year I, willy nilly, organised an eaglet rescue expedition, to one of 'my' Inverness-shire eyries. That year, for my first time ever, when I was nearer fifty years old than forty, I experienced, reluctantly, the sensations a spider must undergo as it slowly spins around, twirls and twists in mid air, on its gossamer-like thread. And that at an age when, to put it kindly, one is assumed to know better, or, less kindly, to be past it.

I had been checking on two eyries in the one day that year, by visiting one, and by checking the other by remote surveillance, i.e. by using my telescope in the evening of that same day. I duly visited the 'personal visit' eyrie which held one thriving eaglet. Eaglets are reared on an exclusively meat diet from infancy and rapidly put on weight on this diet, from two ounces when hatched to ten or ten-and-a-half pounds at ten weeks after hatching. At about ten weeks old they are fully feathered and ready, to all appearance, to fly from the eyrie. Part of this meat diet is often in the shape of fox, most often cubs but on occasion, adults. On this very eyrie I had already recorded two fox cubs as prey. Both bore the typical talon punctures about their midribs, where the fierce clench of enormously strong eagle feet had squeezed the life out of them. As a relative guide to the rapidity of growth in young eagle and young fox, the young eagle at five weeks after hatching was seven-and-a-quarter pounds in weight whereas the fox cub of eight weeks weighed only two-and-a-half pounds.

On my return in the evening I checked up on the other eyrie, the light then being good on it. This was, to me, a quite inaccessible eyrie, secure in its site between its sheer rock buttresses and with an overhang of rock above which prevented even a look into it. A skilled rock-climber could, I had no doubt, get in to it; I have never been an enthusiastic rock-climber, in my more pessimistic moods regarding this as a more or less legitimate way of commiting suicide. Colouring my attitude to this eyrie's rock face was my knowledge that, in the not so remote past, a keeper, in trying to descend into the eyrie from above, had fallen and broken his back. The rope he had been using had frayed through on the rock edge of the overhang. I was to wish that I had never been told of this unfortunate keeper.

Lying comfortably on a heathery bank I spied across at the distant eyrie. With some puzzlement I saw the dark, bulky shape of a female eagle on an inner edge of the eyrie where a narrow crevice intervened between nest structure and rock buttress. What was she doing? She appeared to be on the point of over-balancing, her head and neck thrust down into this crack. Her posture struck a bell; it was the stooping posture adopted when offering scraps of food to a young eaglet. The 'clue' was confirmed when I then glimpsed, with straining eye, through my 20X telescope, a mere scrap of white deep in the crevice. Minutes of concentrated eye-watering scrutiny established that it was a young eaglet and that the female was almost overbalancing in attempting to reach down to feed it. It was also all too obvious that the eaglet was quite unable to regain the eyrie and equally that the female had no idea of how to aid it. To say that I was dismayed is putting it lightly. This eyrie had had only addled eggs for the past three nesting seasons; was this fourth season to end prematurely in failure also? I had tried, year after year, to get a way up the rock face to this eyrie. Frustratingly, I had always failed. It very definitely required skilled rope work and the knowledge of an experienced rock-climber.

There was nothing I could do that night; I still had nearly a hundred miles to drive, to reach Torridon. Once there I would phone Charlie Rose, a friend of mine, who actually enjoyed tackling difficult rock faces. He was, unlike me, a skilled and careful rock-climber. Charlie proved sickeningly enthusiastic about the idea of working out the problem of the climb to rescue the eaglet. Strangely, I slept well until, early next day, it was time to drive to pick up Charlie and to pick up Tom another eagle enthusiast who was eager to help. In June, daylight comes early and we enjoyed the drive, the world around us fresh in the dawn wakening. I kept hoping that we would not need to tackle that fearsome rock face, that the eaglet had somehow become reinstated up on the nest. My elder son, Lea, was waiting for us as we arrived at our departure point to take the route to the eyrie. He had younger eyes than I had, and he had also been able to spot the white object which was the eaglet, still in its crevice. The die was cast; ropes and all sorts of fearsome rock-climbing gear were shared out amongst us and off we set. Weighed down even more heavily by sundry dark forebodings, I led the way towards the fearsome cliff. Even the thought that I would at least be able to see into this hitherto anonymous eyrie did not really cheer me up.

An hour later we stood below the eyrie cliff; Charlie probed here and there, testing rock cracks, peering up at the overhang, and then, rather to my surprise, confirmed my own unskilled diagnosis that it could not be climbed up to with safety. We would have a further steep climb, skirting one edge of the rock crag, to get above the eyrie from where we would have to abseil into the nest. One of the main problems which had confronted me in former years was that, above the rock overhang, the slope was rotten with wet, slippery, mossy growth, without tree or even a stable rock to fasten a rope to. Charlie was not in the least put out by this! He had an assortment of wedges to jam into any rock crack which he judged safe to afford an anchor. He was firm in his belief that these tiny wedges would hold our weight. We scattered, under Charlie's expert direction, to look for suitable bedrock. Way above the rock lip of the overhang Charlie struck what he regarded as 'safe' rock, buried in moss and heather. I would have passed it by as quite unthinkable. We cleared this insignificant reef of rock of heather, moss and, deeper still, moist, black peat. Meanwhile Charlie was sorting out his ironmongery of wedges. Hoping for something akin to a Cunard liner's anchor, I was dismayed anew at the tiny slivers of steel to which we were to trust ourselves in our descent. Charlie, at least a couple of stones heavier than I, had no such qualms. I made a mental vow to myself to 'permit' Charlie to make the descent first; after all, he was the rock enthusiast. The minute cracks we exposed in the rock appeared ridiculously miniscule to the eyes. 'Don't worry' said Charlie, 'we'll test them first.' Blithely he selected two tiny wedges with their attached loops of fragile-appearing wire and, one by one, hammered these in. 'Right', he said, 'test that one', while he heaved at the other. 'Grand!' he said. In utter desperation now, I heaved hard at mine – a moment it resisted, then out it came. Far from being dismayed, I felt that I had been vindicated. 'Oh, well,' said Charlie 'obviously not quite the right angle to hold well.' Quickly, he discovered another crack, hammered a wedge in, heaved strongly and steadily on it and pronounced it safe. All well now, our abseiling rope was attached to the angled 'security' of the two wedge-held steel wire loops. A so-called safety rope was attached (purely as a reassurance, said Charlie) and the appointed hour was on us.

Charlied buckled himself into a web harness, rather as if going parachuting, instructing me at the same time as to the mysteries of this. He clipped onto his wedge-'secured' rope, now dangling over the rock edge, turned his back to the drop then, leaning back, all his weight on the rope, walked back and out of sight. I didn't even have time to say 'cheerio' to him.

Moments later, while I was keeping my eyes on the wedges and my ears open for the crash below, Charlie's voice floated up to me. 'I'm on the eyrie! There is a chick at the bottom of the crevice right enough. There is also a eaglet on the eyrie!' It was my turn now, to re-enact Charlie's spiderman performance. Bearing, or rather attempting to bear his instructions in mind I went apprehensively backwards over the edge, lacking, entirely, the nonchalant aplomb which Charlie had shown. So badly overhung was the crag that my feet lost contact with the rock almost at once. Now, with all my weight harnessed on the rope, I eased my way down, spinning slowly like a grotesque spider, until I was opposite the eyrie ledge. There I hung in what is, very aptly, called thin air while I nerved myself to try to swing myself in over the six or seven foot gap to the eyrie ledge. To my further dismay this was sloping down and outwards. It took me half a dozen increasingly panic-stricken attempts to gain a toe hold on that elusive ledge. Safely there, literally shuddering with relief, I sat on the bulky eyrie and contemplated the two eaglets, lying side by side. The one which had caused my initiation into abseiling was the smaller of the two and had one white down-clad wing joint rubbed red raw, in its effort to regain the nest. Overhead, high in the air, was the female, now visible, now hidden, by the overhanging bulge of rock above me. The eyrie ledge was snug and sheltered but for its outer edge.

I photographed the eaglets and tried to be more nonchalant in appearance this time as I again walked backwards over the eyrie edge to abseil to my helpers below. I believe that they felt some relief in seeing my amateurish attempt at abseiling accomplished successfully. I assure you that it was nothing to the relief which I felt.

The long walk back to the car was accomplished to the clink of climbing ironmongery and the clack of tension-released tongues. I shall forever be grateful to Charlie Rose, especially for his ready acquiescence to my appeal for help. Without him, our efforts would have been fruitless. Tom Wallace, my son Lea, and myself, were willing enough but it was Charlie who supplied the equipment, the expertise and the cool competence. Above all, he kept our morale high.

1980 proved to be a bad year for golden eagles in the Highlands. Those obsessive, self-centred characters, the egg collectors, more particularly those who battened onto the eggs of rare and endangered birds, were pillaging eyries throughout the North. Like those other predatory species, fox and hoodie crow, but with less worthy motives, these unprincipled scoundrels work in pairs. On 25 March 1980 a pair of these egg stealers was apprehended and found to be in possession of two clutches of golden eagle's eggs. As a result of a tip-off the RSPB had been maintaining a watch on the two men concerned and they knew the eyries from which the four eggs came. In previous years I had co-operated closely with the RSPB and more than once had suggested to their representative in the North that, when eggs were recovered quickly from eyrie robbers, the slim chance should be taken of returning them to their eyries. This suggestion came to fruition in 1980 since the eggs stolen must have been recently laid.

Because one of the eyries robbed was one of my study pairs I was asked by the RSPB to help once more by continuing to monitor this nest, in which the stolen eggs had been replaced, after an absence of twenty-four hours. I was more than willing; I had been watching this particular pair since 1963 and I knew the female very well indeed. She was getting old now but she was an exceptionally bold character, devoted to her young and to their protection. I was absolutely convinced that she would return to her eyrie despite the twenty-four hours' absence of her eggs. The unknown factor was whether the fertility of the returned eggs had been impaired by their short absence from the eyrie. Since the hatching time for the golden eagle is around the first week in May (the eagle has a very long incubation period of about forty-three days) I initially decided to forbear from a visit to the eyrie until then. My interest and anxiety was so great however that I found that I just could not wait until then. The dawn of 27 April saw me on the rugged, testing journey through the hills to the eyrie. As always, deer, shabby looking as their thick coat of winter became displaced, were everywhere on that hill journey. While I took a breather, on attaining a wide-ranging plateau of 2000 feet in altitude, I took extra time to count the scattered groups which were still visible. There were more than one hundred deer grazing avidly in the mild spring air. Golden plover I knew were present, hearing their plaintive whistle; grouse I saw and was exhorted by a cock grouse to *go-bac, go-bac*. A greenshank I also saw, early on her usual nesting ridge, flitting away on flickering wings. At one point I was aware of a ghostly, grey-white shape loping away, silently from a cluster of rocks. When far enough distant this apparition rose upright, on its hind legs, to turn an enquiring long-eared, nostril-quivering head in my direction. No wraith, just a mountain hare. A pair of mountain blackbirds (ring ousels) I saw and also a pair of dumpy-of-body dippers by a burn waterfall. All these I noted, almost absent mindedly; they were all part of the predictable hill scenario but my mind was centred, that day, on the eyrie which I was now nearing.

Coming into the actual glen of the eyrie almost the first thing I saw was the huge hulking shape of a golden eagle perched on an even more hulking rock. Its back was to me and momentarily I admired the gleam of the pale golden head in the early morning sun. Only a moment I had; the eagle, somehow sensed my appearance, a scant hundred yards away and, unhurriedly, launched itself with a powerful shove, off its rock. Encouraged, rejuvenated, I pressed on the half mile or so to be opposite the eyrie. Sitting down, I spied with my invaluable stalking glass at the bulky structure. With joy I saw the pale head of the female showing above its rim as she incubated, deep sitting, in the centre cup of the nest. Now, as I have said, I knew that this eagle was a bold, courageous bird, nevertheless it came as an enormous relief to know that she had justified my belief in her. The initial tremendous hurdle of her acceptance of her restored eggs was over. I looked forward to early May with well nigh unbearable anticipation.

It was to be 10 May before I was again afoot, at dawn, in pale, strengthening sunlight, on my way to this eyrie. Again it proved to be a blessedly dry day, mist-free, so that one revelled in the sights and sounds of the year-freshening month of May. The air was decidedly bracing, a cool south-easterly wind blowing off the few snow wreaths which yet patched the higher ridges. That day I saw no deer while crossing the high plateau but the greenshank pair, near a small lochan, were much more vociferous than on my previous visit. I almost stepped on top of a pair of equally voluble grouse, and a golden plover flitted ahead of me, whistling mournfully. Toiling along on the steep face across the glen from the eyrie, I had the usual thrill when an eagle appeared above me, to fly ahead to perch on the steep rock outcrop which was the guard post opposite the eyrie. I sat down and spied across at the now visible, if distant, nest. As I'd suspected, there was no adult eagle sitting on it; it was the female eagle who had flown over me. Welcoming me to her glen? More likely warning that she was watching me!

It was in a rather pessimistic mood that I descended into the glen and then tackled the steep rock-strewn face up to the eyrie. Maybe the two eggs had failed to hatch after all! Why was the female eagle not sitting tight on her nest?

The actual cup in the centre of the huge nest was so deep that one could not actually see what it contained until one was standing at the nest edge. I could see however that a dead, headless rabbit lay on the widest rim, completely denuded of fur. Hope sprung anew to banish my pessimism. An eager last few careful steps and I was perched at the edge looking down at one small, down-clad, days-old eaglet, pink skin showing through the pure white fluffy down which clothed it. Undisturbed by my arrival it lay asleep, cuddled into the unhatched egg which also lay there. I picked up this egg and gently shook it. A sloshing, liquid sound told me that it was addled. As I stood there, bemused and happy, in a sort of joyful daze, a tremendous 'swoosh' of displaced air from giant wings aroused me. My female eagle was 'buzzing' me and was even now banking around above, to sweep at me again. Yet again, (as in every year since 1963 and this was 1980) I was very thankful for the deep sheltering overhang which prevented her closer approach. I had come to cultivate a strong affection for this selflessly aggressive female eagle, so protective of her young, but I had no wish to have any type of close quarter hand-to-talon contact with her. She repeated her swooping tactics half a dozen times while I photographed her eaglet. Beside the fur-denuded rabbit lay a dead black water vole, or rather it had been black before all its fur had been plucked out, leaving it starkly, and rather obscenely, pink; its curved orange-coloured incisor teeth yet clenched tight on a blade of green grass. I left and, when halfway up the opposite face, looked back in time to see the female alight on the eyrie with that characteristic upward and backward fling of giant wings and out-thrust talons. A moment she stood peering down at the eaglet, then, awkward-looking but infinitely careful, shuffled forward to brood her eaglet.

I monitored the progress of the 'rescued' eaglet weekly from that day onward, including weighing it. You may speculate as to 'how' this weekly weighing was accomplished. Simple enough really, when the eaglet was small, but it became increasingly difficult as the eaglet grew in size, in weight and in strength. One used, at first, a plastic bag; the eaglet was popped into this, the hook of a spring balance inserted and there you were, reading off the weight. At a weight of two ounces on hatching the process was easy and painless. At an age of nine to ten weeks and a weight of nine or ten pounds it was rather more difficult and, at times, painful. The plastic bag used had to grow progressively larger and stronger; they became expendable. After each weighing of the eaglet, with head and beak designedly out of one end and at least one wildly-striking, steel-hooked foot out, not designedly, at the other end of the bag it was not even remotely usable thereafter.

This eaglet seemed to me to be a particularly lively and precocious one, a male, I thought, since, at nine weeks hatched, he weighed a bare nine pounds. At eight weeks after hatching he succeeded in grabbing the fingers of my right hand in such a pulverising talon grip that it hurt excruciatingly. It took me many pain-filled moments to prise the talons loose, for the eagle's reflexes cause the talons to contract and so sink them in more deeply when a victim struggles. The next day the middle finger of my right hand, into which one talon had sunk deep, was stiff, swollen and painful, I had rather disquieting misgivings concerning infection for a few days. Luckily, or perhaps logically, since I was handling 'wild' creatures regularly, I had kept up regular anti-tetanus injections. My finger had healed sufficiently for me to be able to weigh the eaglet upon my next visit. It occurred to me that the boldness of this eaglet was probably a legacy from his mother. She, as I have said, was the boldest eagle I have ever encountered in a long innings in studying the species. She seemed also to be able to count, for she never ever attacked me while I had a companion with me as, occasionally, I did have either a friend or one of my sons. When I was alone, however, at her eyrie, she invariably attacked me. One would almost have thought that, in that year, she knew that her eyrie had been under threat. Perhaps it was nothing more significant than that she was getting older, and bolder. Certainly this proved to be her last breeding season and I was to be happy that she had ended on a high note.

In any event, she was bolder and even more persistent in her attacks than ever before. Fortunately, there is a deep rock overhang over this eyrie which shelters it superlatively from the attacks of the protective female eagle. This rock overhang juts out sufficiently to mask out much of the hill to either side of the eyrie also, so that, crouched on the nest, photographing or weighing the eaglet, I had no view, no warning whatever, of that last, fast, low glide, talons down, of the female attacking. It became rather nerve-wrackingly apparent that the attacking eagle knew precisely the height and extent of that overhang. It also became apparent, even more nerve-wrackingly so, that she appeared to be timing her attacks to coincide with my rapt concentration on her eaglet. Time after time, on each visit I made to the eyrie, she caught me completely off guard, swooping so silently, so close and so incredibly fast, that there were times when she even caused the eaglet to flinch, perceptibly. This was the first time, too, for all her years of swooping at me, that I had heard a wingtip, the inner wingtip, actually rasp on the rock face above my crouched figure. I managed to see that it was on the upward stroke of this huge wing that this rasping occurred and, in time, I grew accustomed to it. It made me reflect on, and respect, the manoeuvrability of this bird, with her wing span of seven feet, that she could venture so close to injury as to make contact with one wingtip, on an implacably hard rock face. I must admit that I did take to wearing a hat on my visits; I could not afford to lose any more hair from my head. I will also admit that I'd have been better pleased if I still had had the steel helmet of my Army days. Perhaps the most psychologically disquieting element was the limited vision I had of the imminence of an attack. The attacking eagle would suddenly appear at the edge of the outcrop, gigantic, dark, relentless, gleaming eyes close-set each side of the jutting beak which, prow-like, seemed lined up level with my shrinking form. Then would come that eerie rasp as the inner wingtip fouled the rock and, in so doing, kept her from actual contact. Out of sight she would go, to reappear high above, gaining height, impetus and angle for her next swoop. I grew blasé; as long as I was sufficiently under the overhang she just could not hit me without damage to herself. Perhaps I grew too blasé for, on one never to be forgotten occasion, I had forgotten to get sufficiently under the overhang and, to get the angle I wanted to photograph the now well-grown eaglet, I was crouched on the outer edge of the eyrie, by now just a flattened, rather tawdry-looking, platform. Engrossed thus my rump, the highest part of me, stuck up in the air, I suddenly received a hard, sharp rap on this, immediately followed by the nerve-tingling 'swoosh' as my now not-so-favourite eagle pulled out of her swoop. To this day I do not know whether it was a wing-tip or a closed-knuckle foot. I suspect the latter, for I had a tear in the loose skirt of my jacket to show for it. Do you know, I was blasé enough to maintain my position, for I still had not got my photo. Her next pass and she hit my elbow, poked outwards as I focussed my camera. Imagine the reaction of the audience, at your camera club, as you unblushingly told them, you had had camera shake as the result of an eagle hitting you? Yet, so it was! It was to be many years later before Dave Dick, working for the RSPB, obtained corroborative evidence of this sort of thing personally when, at an eyrie he was checking, an eagle made 'contact' with him and left him with a bleeding scalp!

Prey at my eyrie followed the pattern I had grown accustomed to over the years; rabbit, mountain hare, grouse, ptarmigan, water vole, fox cub and meadow pipit nestlings. I recorded no deer calf nor lamb in that nesting season though I had done so in other seasons in that same glen.

The eaglet left the eyrie on 12 July, walking out of it, I thought, since I had seen this happen before at this particular site, from which the eaglet did not have to fly in order to be able to leave. I could not locate it, only the remains of a dead fox cub, on the rough hillside above the eyrie. I did, however, hear its ridiculously juvenile cheep-cheep-cheep food call as I left the glen. I wished him luck.

That year, 1980, was a busy year for me for I was also visiting two other eyries in Torridon. I had a tremendous thrill at one of these, the more so since it was entirely unexpected. I had got to the eyrie very early one morning and was in position, watching, well screened by a rowan tree growing in the gully close to the eyrie. Suddenly the eaglet gave tongue in a clamorous, excited kewping and before this had fully registered in my mind the male eagle swept in. He landed, oh so neatly, with a flourish of wings, on the edge of the eyrie, despite having the weight (around five-and-a-half pounds) of a mountain hare clutched in one foot. Burnished gold he seemed, gleaming in the early sunlight, little wonder Imperial Rome chose the eagle as a standard. Only a brief moment he rested there, looking down at the eaglet, now mantling over the prey he had brought in, before he flung off the eyrie, dropping slightly, until the unfolding wings took his weight. Scarce had he gone when the female arrived, much bulkier and darker of plumage than her mate. She too paused momentarily, contemplating the hungry and now near frantic eaglet who was cheeping stridently in his lust for food. She went to work then, standing over the hare, pinning it down in her two powerful feet and wrenching pieces of flesh, wisps of fur and bones and bits of sinew, all dangling, raw and bloody, from her hooked beak. Fierce, elemental, raw, as this tearing up of the prey was, there was a very notable aura of tenderness about the huge raptor as she proferred, with infinite patience and gentleness, tiny scraps of flesh to her offspring. If too large a piece was wrenched off by the impressive beak she swallowed this herself. I watched her part, delicately, the long thigh bone of the hare from its lower joint and, almost devoid of flesh as it was, attempt to swallow this. She had it halfway down when the over-ambitious and ravenous eaglet seized the protruding end and actually pulled it out of her unresisting beak. If it was difficult for the mother to swallow it was quite impossible for the eaglet. The female watched, patiently, until she judged that enough was enough, when she retrieved it and this time swallowed it. I believe that a predator, such as the eagle, without benefit of saw, or scalpel, is just as proficient as any surgeon in its knowledge of anatomy and hence capability in dissection with its beak. A red deer calf, for instance, too heavy to be carried off in sustained flight, will be split midway along the spine, at a joint.

Potential prey species, or perhaps I should qualify this by saying the young of potential prey species, seemed in no awe of the proximity of an occupied eagle eyrie. I have often noted sheep and lambs grazing or lying below such an eyrie, and similarly, red deer hinds and their calves. On one occasion while at an eyrie I watched and wondered as three beautifully dappled and appealing red deer calves trotted, highstepping, through a gully below, all by themselves, not a hind to be seen anywhere. Full of *joie de vivre* they were too, prancing and pirouetting, springing to one side on stiffened legs, playing 'king of the castle' on any inviting knoll. Noisy they were also, bleating to each other quite unafraid as they went their way with absolute impunity. It was a delight to watch in that early July sunlight yet I was forebodingly apprehensive that that innocent frolicking would be rudely terminated by the thud of an eagle as it fixed its talons into an all-too-tempting feast. Small birds had no inbuilt fear either. I have seen even the tiny willow warbler hunting flies, in the foliage overhanging an occupied eyrie. Visits of the brown wren to an occupied eyrie were commonplace also. I suppose it is logical enough; prey remains, infinitessimal as they might be, attracted insect life, therefore an eyrie could be rich hunting for an insectivorous bird. I remember being exceedingly entertained one day by a perky wren busily hunting insects under my nose as it were. On the further edge of the wide platform of the eyrie was crouched the eaglet, hungry, its crop empty. On the near edge the wren flitted busily. Incongruous yet vastly amusing, it was 'confrontation' between the hulking, hungry, glowering eaglet and the tiny, neat, head-bobbing, tail-flicking, wren, no whit abashed by the glare of the eaglet. Hunting over, the wren left, to hunt in the rock crannies above the nest – a lion and the lamb situation, if ever there was.

I had long wanted to spend all night in a hide at an eagle's eyrie and on a calm, warm evening one May I yielded to that urge. It was 6pm when I left on the long rough trek to the occupied eyrie, burdened like a pack mule, with hide, sleeping bag, provisions and, of course, my camera. Burdened like this, I needed quite a few rests before I reached the eyrie around an hour and a half later. On arrival I eased off my load and, tired as I was, set to improvising a hide straightaway. I had seen no sight of either adult and I wanted to get hidden before one came to the eyrie.

I got my hide up very quickly; it was fairly sketchy and designed for concealment rather than comfort. By 8 pm I was installed; space was limited and my rucksack took up fully half the 'floorspace' while my rolled-out sleeping bag extended right to the edge of a nasty vertical drop. I waited, tense with that almost unbearable sense of anticipation which concealment near to an eagle's eyrie inevitably aroused in me. Nor was my wait of long duration for, within ten minutes, the female arrived, took a reassuring look at her dozing eaglet and left almost immediately. It is worth noting that in all her comings and goings she was absolutely silent, as silent as any night-wafting barn owl so that I never ever had any audible warning of her arrival. Silently, ghostly silent, she would arrive, huge, on an outer edge of the eyrie.

By 8.30 pm the sun on that lovely tranquil May evening was shining low, along the long narrow sea loch visible from my perch. The water was a huge pathway of liquid gold so brilliant that it hurt the eyes to drink it in for too long. This particular eyrie had a magnificent view spread out below its rocky height. A rock-strewn hill face fell steeply to the glen far below with its silvery burn winding a sinuous way to the far-off sun-glittering loch. The low sun, sinking ever lower, silhouetted a chain of jagged peaks, black and distant, on the horizon, against a calm, lemon-yellow sky.

Amid all this remote tranquillity the female came back twice, on one occasion with a sprig of rowan in her beak which she arranged to her fastidious liking on the rim of the eyrie. She was of course extremely keen sighted; even a very slight movement of my camera's telephoto lens, concealed as it was, caught her eye. A hard-eyed stare at the black 'eye' of it for a moment, then she would resume her fussing around at the eyrie. The sun finally disappeared below the western rim of the mountains about 9.30 pm. Five minutes later the female arrived once more and perched on the outer edge of the eyrie. I tried no more photography; it was time for the female to settle at her nest. I, 'settled' in mine, tried to remain motionless as she now was, watching her, with bated breath, a wild, quite untamed eagle, more at ease in these darkening, wild surroundings than I was, as the grey dusk drew in.

Around 9.45 pm the eaglet, obviously now feeling the night air becoming chilly, sought comfort from the female by making its laborious way from the rock-shielded inner edge of the eyrie to her outer edge. Laborious because, young as it was, its legs were not yet able to bear adequately its body and it 'oared' its way over the eyrie using its tiny flipper-like wings to aid it. The mother, frigidly Spartan in her outlook, ignored it completely; the eaglet, finding the exposed outer edge even colder than its sun-warmed-rock inner niche, retreated thence, cheeping fretfully.

For the next fifteen minutes the female was motionless, a graven statue in buffs and browns, then, about 10 pm, she very, very slowly and carefully edged her way forwards onto the body of the nest, as if in apprehension lest she tread on her eaglet. This brought immediate reaction from the fretfully dozing youngster. The white head came up and it jerked into learning-to-walk activity, making a slow, flipper-balanced struggle towards the female. This time she welcomed it, her golden head with its powerful beak lowered to caress momentarily the tiny white-down-clad chick which in a matter of weeks would be transformed almost to her own impressive stature.

Although the sun was now completely hidden there was yet sufficient refulgent afterglow to be able to watch by its light. To the muted beauty of that superlatively lovely and tranquil evening the eagle, hardly surprisingly, showed no appreciation; her head was turned in to the darkening rock face while she crouched over her now silent and comforted chick. She had seen plenty of such views. That ability of the eagle to remain absolutely immobile for interminable periods never fails to intrigue me; even in digesting a full meal and an adult will spend five or six hours in motionless meditation. Little wonder one seldom spots an eagle unless it is airborne.

As 11 o'clock came it was too dark to see the female distinctly though I could make out that she had still not fully settled to the brooding of the eaglet. I was certain however that she was more comfortable than I was. Throughout that night, half-sitting, half-lying, boots and clothing all on, in my sleeping bag, I could only doze fitfully. The ground became increasingly hard and undiscovered bumps and rocks became, belatedly, discovered. My booted feet began to itch where they braced my semi-reclining form, wedged into the bottom of my hide. Had I been able to stretch myself comfortably these same boots would have stuck out over the drop. Perhaps that was why I could only doze fitfully, my subconscious was aware of that drop. At least I was drowsily warm; my choice of night was a good one.

When, in the first faint glimmerings of dawn I peered at my watch it proved to be 4.30 am. Darkness, in May in the Highlands, is shortlived. The eagle was sound asleep on her eyrie, her eaglet snugly below her breast, quite unseen. Her tail feathers were towards me, her golden head invisible, tucked below a wing. Almost an hour later, at full daylight, she woke; her head came from below her wing and she peered intently downhill to where a ring ousel was saluting the new day. By 6 am the sun was beginning to touch, rosily, a crag to the southwest; the air was so still that there was a tangible silence; now and again a light breeze brought the sound of a waterfall on the burn below, to my ears. I was, it seemed, isolated in a tranquil world of my own, with only the eagle, immobile again, on her nest, near to me. My booted feet began to itch again, reminding me that I was human.

At 6.20 am the brooding body of the eagle seemed to be heaved abruptly upwards by a movement of the eaglet. At once she shifted position, rose to her feet on the edge of the nest and after a lazy survey of the glen below, she flung herself off, unfolding her wings simultaneously. The eaglet, exposed now, lay still for another ten minutes when it half-rose, manoeuvred awkwardly round on the nest until its rear was pointing in the desired direction then sent a tiny jet of white into the void. Early sanitary training for such a young eagle, but necessary when such a long habitation of the eyrie lay ahead before it fledged. I have never ever seen an eagle's eyrie in such a foul, insanitary, smelly mess as that of the peregrine falcon, which has a much shorter fledging period, and obviously, no sense of smell!

I was weary but certainly not bored. I could not even try to doze in case I would miss a visit from either of the eagle parents. I elected to stay put for as long as I could and had an *al fresco* breakfast of sandwiches and an apple.

Time passed; it was near noon before the sleeping eaglet stirred; the day remained pleasantly warm; there was no need for the female eagle to protect her young from inclement weather or from too hot a sun. Shortly after noon she did come in; a shadow flicked past my hide and, noiseless as ever in her approach, the female swooped up on to the nest edge. She stayed only a couple of minutes inspecting the eaglet which was again sleeping, and probably looking for any prey brought in by the male. Disappointed in this, she flung off her eyrie again. Awakened by her thrust from off the nest the eaglet began belatedly to cheep hungrily. A light smirr of rain then seemed to stir the eaglet into activity but it was only to evacuate over the nest edge again.

The female paid another brief visit at 1.25 pm. Again she found no prey and, irritated by this, it seemed, she gave vent to a surprisingly musical bell-like muted *klonk-klonk-klonk* call, followed closely by a more high-pitched *klink* call, querulous, to my imagination, before she flew again. This was the first time I had ever heard the eagle, a notoriously silent bird when adult, utter such a call. The reintroduced white-tailed sea eagle is much more voluble, indeed garrulous, at times. The disappointed eaglet also resumed its somnolence. The breeze was strengthening now; the leaves of the small rowan tree at my hide were fluttering perceptibly. Yet the leaves of the rowan tree which was growing out from the actual eyrie ledge were quite motionless. I began to reflect on how ideally most eyries were situated, and of how the deep, warmly-lined, thickly buttressed, nest cup in which the young were hatched, retained warmth at that early stage when the young needed this most. Bear in mind that most eagle eyries inland in the Highlands are situated at more than a 1000 feet altitude and that on many occasions an eagle has been noted incubating, or brooding, with snow all around. An eagle's eyrie, to be successful, must be so situated so as to take whatever vile weather a Highland spring can throw at it. This is why the overhead protection of a deep overhang is usually such an adjunct of the Highland eyrie. Having said all that, it has become abundantly clear to me that the golden eagle can bear, with equanimity, tremendously cold, hostile weather much better than it can bear a spell of hot cloudless weather. Such hot weather can have the eagle exhibit extreme discomfort, a large bead of moisture rolling intermittently off the point of its beak while it pants, at intervals, beak gaping open. Any bit of shade is sought, making a perched eagle, in deep shadow, even more difficult to spot. Wings and legs will be periodically stretched; the entire picture is one of uncomfortable distress.

Soon after 4.30 pm I left my improvised hide. The halcyon weather which had made my vigil so enjoyable had, in a sense, been against me. The eaglet did not need brooding; by its evacuations it had been well fed before my vigil commenced. I had had twenty-one-and-a-half hours of watching and I was happy. The female, at this early stage was obviously depending exclusively on her mate to do all the hunting. She was maintaining the equally essential task of watching over the eyrie – her eaglet was still very vulnerable.

The eaglet was reared successfully – it left the eyrie, fully fledged, on 20 July.

It was at this eyrie that I actually watched the eagle bring in a dead fledgling kestrel, clutched tight in one talon, while she was hotly pursued by the mother kestrel. On 23 June I was about to leave the eyrie and was standing motionless, screened by the overhanging cliff and also shielded by a rowan tree. Suddenly, so close that I could have reached out and caught a wingtip as she passed, the eagle glided alongside, parallel to the cliff, with a small object held enclosed in one huge foot. So intent had she to be, on her close-to-the-cliff nagivation, that she did not see my motionless figure. It was a quite magnificent close-up view of the gliding eagle and hard on the tail of the fast-gliding giant was a diminuitive female kestrel, fast flicking narrow wings just a blur. She was calling incessantly in heartbreaking distress. It was not the strident, harsh *kek-kek-kek* of alarm but a pathetic low-pitched, reiterated call, eloquent, oh so eloquent, in its anguish. Bravely, persistently, but, of course, unavailingly, she rode on the eagle's tail until the eagle swooped up, unerringly, to land on the edge of her eyrie. Then, tiny, she turned away, protesting calls quieted, silhouetted, translucent, against the strong sun, in a red brown haze, of flickering wings and black-barred tail.

One, of course, as an observer of wildlife, should never get emotionally involved, yet I must admit that my sympathies that day were all with the tiny, courageous mother kestrel.

Over my years of eagle watching I formed the opinion that the golden eagle is a magnificent and imperious appearing bird but one which, probably luckily, and certainly with reason, is intimidated by mankind. It is a dramatic bird of prey fitting in well to its niche in mountainous scenery; it does not appear to me to be a very intelligent bird; capable yes, courageous yes, but most of all, adaptable. It is capable of tackling any of our present day wild species, four legged or feathered, which exist in its sphere of operations, and, on occasion, does so – though not always with success. Red deer have killed a grounded eagle – a hind or hinds, trampling an eagle literally, into the peat, an eagle which could not rise from her attack on the hind's calf quickly enough. There is an instance on record where a golden eagle and a wildcat, in a fight over prey, both died, the wildcat by being dropped from a height; the eagle, found later, was dying from injuries inflicted by the wildcat in the struggle. I know of one instance, and one instance only, of a dead eagle being found at a fox den. Now the fox will pick up carrion – that particular eagle may well have already been dead when the opportunistic fox found it and dragged it to its den. Nevertheless these instances will show that an eagle does not always win, in taking risks, to obtain its prey.

The most striking animals that I have found as prey of the eagle have all been predators, fox, stoat, weasel; wildcat kittens, badger cub, feral mink and pine marten have been recorded by other observers. The most amazing prey ever was that of a peregrine falcon; I found this on two occasions, in different years, at widely separate eyries.

I found that the range of prey taken in by a pair of eagles was amazingly large but that the staple was undoubtedly mountain hare. Once the female eagle of a pair ceased concentrating all her energies on guarding her eyrie she was capable of catching larger and heavier prey than the smaller male. He on the other hand being smaller and perhaps more agile on the wing could catch the more nimble prey which might be able to out-manoeuvre the bulkier female eagle. Most of the live prey eagles caught were, I believe, taken from off the ground, in a silent swoop. The technique seemed perfectly adequate if less spectacular than that of peregrine falcon or even sparrowhawk.

Added to their almost unlimited live prey capabilities is the fact that the eagle, whether it spoils your romantic picture of it or not, does not disdain carrion; the eating of dead deer, dead sheep, dead fox or even, as I found once, dead horse, gives it an extra dimension in its adaptability which is the key to its survival.

The eagle, for all its impressive size and undoubted powers is, perhaps wisely, certainly logically, wary of man. In the case of the larger female, she it is who always keeps watch on her eyrie when the eaglet is young, and hence is vulnerable, while the male hunts for prey. She will use a 'guard post' in such a position that she is within easy range of her eyrie with an all round view of it and its environs. With the eagle's superb vision this may well be on the opposite side of a narrow glen, a long trudge for humans or for four legged predators but only seconds' flying time for the sentinel eagle. This constant habit of the eagle to watch her eyrie is, it seems to me, more as a safeguard against predators which may inhabit the same glen. Many eagle eyries are accessible to both fox and pine marten and both are agile and opportunistic hunters. Remember that a recently-hatched eaglet may weigh only two ounces which is easy prey for a hungry stoat, or even a weasel. I have seen hoofmarks and droppings of both deer and wild goat on an eagle's eyrie ledge. I once discovered the cast antler of a young red deer stag caught in the structure of an eagle's eyrie and a stalker colleague once found a dead stag dangling by a hind leg from one of the old soup plate-sized eagle traps which were regularly used in the not so distant past. Hard luck on the stag you may feel – poetic justice perhaps on the stalker who set the trap. The pine marten, I have found, uses certain eyries of the golden eagle as temporary resting places, or as more permanent holing up places, when, obviously, these eyries are not, that year, in use by the eagle. In the former case, I have found a comfy couch actually made of hair from a dead deer carcase, on the surface of the disused eyrie. A well-used marten latrine was at one inner edge of this eyrie. In the latter case a hole will be burrowed out, deep into the actual side structure of the eyrie; to excavate such a tunnel, deep and cosy, into the compacted structure of an eyrie, is well within the powers of the versatile pine marten. Seldom does the huge eagle make any overt demonstration should a human being appear at an eyrie; she prefers to watch from a discreet distance, more often than not her vigil is quite unsuspected. Is this a legacy of nearly a century of unchecked persecution, from the days when any visitor to an eyrie probably carried a gun and was usually intent on using this. If there were bolder eagles in those days they would have been shot, time after time, while the more wary survived, to foster, eventually, a more wary species.

On the other hand I have never known a golden eagle to desert its eyrie because of human disturbance, after it had hatched its eggs. There are birds, such as the common buzzard, which are notoriously prone to desert their nest at the slightest disturbance. The golden eagle is not such a bird, which of course, does not give licence to disturb such a nest without due reason. It is illegal to do this in any case; the welfare of our eagle population must remain paramount. The few recorded cases of eagle aggression (or protection, if you like) on a human interloper at an eyrie may increase, as bolder eagles, nowadays, will escape a charge of shot. I may add that, having first experienced an attack on myself, by a protective female early in 1959, I, thereafter, was always wary, until I had established the particular female's attitude. One should always be aware of the potential of a possible adversary. The potential of eagle talons, steel-like black hooks, needle sharp, with their spread of six inches and their destructive, constricting, convulsive grip, tightening instinctively by reflex, not slackening, as one struggles is lethal. An eagle, by preference, seems to prefer to attack from the rear; the back of a human neck is singularly unprotected from a fast, silent, aerial swoop by a projectile weighing nine to ten pounds, especially when the human is precariously perched, ascending to an eyrie.

12. Swansong

I have often been asked if I miss Torridon or why I did not retire to a house at Torridon. Of course, I miss many aspects of Torridon; one cannot live for almost twenty-two years in such an area and not miss it. But then you know I had been born and brought up in Fort Augustus and all my early formative associations were with that Inverness-shire village. In all my years at Torridon I never forgot those youthful and happy associations and I never swerved from my resolve to return there, when I retired.

My memories were of a happy if rather turbulent childhood in the then largely pastoral village of Fort Augustus. Turbulent, I should explain, only because I am told that I was 'an imp of the devil' and unlikely to survive to adulthood. As to this particular accusation I cannot, of course, testify, although, if continual adult assurances were to be believed, I stand convicted.

I remember that there were few cars in the village; before I was of school age I was once taken through the village happily astride the boney back of a knobbly-spined, wide-horned, Ayrshire cow, one of a dozen being herded home to be milked by old John Smith, who at that time had the tenancy of Loch Unagan farm. Not a car passed us, to disturb my steed.

Instead of a road as main thoroughfare to Inverness the Caledonian Canal then served that purpose. The old broad-of-girth paddle steamer, the *Gondolier*, was a familiar and welcome sight to us, going to and from Inverness and Fort William. We children, according to our behaviour on the day in question, used to get a ride the lengths of the locks on board her. On privileged occasions we were even allowed into the warm engine room, with its, to us, massive gleaming, brass rods and cylinders and its all-pervading smell of hot oil. I developed, most ill-timedly, badly septic tonsils one evening long after the *Gondolier* had gone, white-waked, to Inverness. I was 'rushed' into the Northern Infirmary in Inverness by car – the doctor's car, a wide-bodied old Austin. The road along which we had to drive was on the south side of Loch Ness (the existing road was not yet made), via Stratherrick. This road was very narrow, so much so that I have vivid memories of our car and the only other car which we met on that journey, becoming jammed together while trying to pass each other. Wrapped in a blanket, held by my mother in the back seat, I felt too sick to really enjoy our small drama.

There was little electricity throughout the Highlands then; the Abbey had its own private turbine and turbine house to generate lighting and lighting only. Some of the bigger houses in the village were connected, again for lighting only, with this supply. Ours happened to be one of these yet our smaller rooms including our bedroom were not connected. I remember vividly the stink of hot paraffin in the black-painted cylindrical Valor heaters; the glass chimneys of the paraffin lamps, liable to sooty blackness if the wick was badly trimmed. I also remember the flickering candle, odorous with melting candle wax. Just one of the juvenile escapades which lent colour to my *enfant terrible* reputation took place because of our candle-lit state. Our bedroom, as very junior members of the household, was directly off the scullery which had a stone floor. This floor, in the pitch darkness of night was a playground for some large, black, hard-shelled cockroaches which smelled pungently when squashed flat by an immensely satisfying skelp with a slipper or anything similar handy.

In the pitch darkness of a long night I was always terrified of the distinct possibility of one of these cockroaches seeking the warmth of the bed I shared with my two brothers. So much so that one night, against all rules, regulations and common sense, I contrived to sneak an absolutely full, new box of matches to bed with me. Sure enough, I woke in the Stygian dark, everyone else around me fast alseep, to feel the distinct tickle of a beetle crawling over me. I cannot remember whether I howled in an appeal for aid or not – I can remember that when, with panicky, sleepy hands I struck a match, that the entire contents of that full matchbox exploded into eyes-dazzling flame. The pyrotechnics and not least the consequent and exceedingly understandable adult outrage which ensued must have scared every cockroach out of the house.

The favourite perch of my brothers, Hugh and David, and myself, was on the corrugated iron roof of our coal shed which, conveniently, overlooked the minister's vegetable garden. Mr Kesting, we knew from personal experience, grew tasty young vegetables; we made sporadic raids on his tender young carrots and his rows of peas after he had gone in for his tea, as much for the dare as because we enjoyed vegetables. The churchyard also lured us because it held some large and stately chestnut trees; it was from one of these that I had my only childhood fall out of a tree. That fall frightened me terribly; I landed flat on my back, arms and legs clutching space, and knocking every breath out of my body. That agony, of uncertainty, of striving to achieve the hitherto taken-for-granted gift of breathing, from a completely winded body, I can visualise even now. I'm afraid it didn't stop me from climbing trees; the next major ascent was of a towering Douglas fir at the bottom of our big garden. Very unfortunately this tree was in view from a house window and I was yet only half way up when I had two of my aunties at the bottom of the tree calling me to come down. Hard-hearted little monster that I was, I carried on. Calls became pleas, then tearful implorings, for my aunties were, strangely, very fond of the three of us. I did, however, attain that coveted tree-top and got down again unscathed. Stirring days, those were.

Going to school, about a mile's walk away, to the handsome building which still stands, up Bunoich brae, was a new adventure. We were often late, catching tiddlers, tadpoles or newts, whatever offered, with glass jam jars in the wee burn beside our road. Or lingering, on warm days in summer, testing the squashy warmth, with bare, big toes, of the soft bubbles of tar which rose then on the road. The smiddy, also on our way to school, was another seldom-resisted lure, hot sparks perpetually flying from the glowing red forge, its long bellows-handle always pleading to be pumped. The ring of hammer, on horse's shoe, on the squat anvil, is evocative even to this day when smiddies are scarce. We were forbidden, quite fruitlessly, to walk to school via the towering viaduct which crossed the River Oich, carrying the railway track to its terminal in Loch Ness, now disused. We were expected instead to use the wooden bridge which crossed the river Oich at Inveroich, near its entry into Loch Ness. Both routes had their attractions for us; the viaduct because it was forbidden and because of the sensation engendered by the awesome drop to the river far below. The wooden bridge at Inveroich because it went by the smiddy and our piscatorial-pursuits burn. Adders and polecats were conjured up by devious adults as hazards to encounter if we went by the viaduct route. We did not even know then what a polecat was like but the chance of seeing a live adder was to us an attraction not a deterrent. We never did see either of these species though there were two or three dead adders always to be seen, eternally immured in bottles of methylated spirits on display in Robbie Cameron's shop on the canal side.

The squeaking of slate pencils, scratching of letters and figures on the school slates, is another evocative sound, as is the rattling of coloured beads on a counting frame. The headmaster (called 'Spratty' by us pupils, but not to his face) was a figure of doom to us primary pupils, aloof, remote: he it was to whom was entrusted the task of using 'the strap', a leather instrument of punishment which we feared and respected. Despite my juvenile reputation I did not often experience 'the strap'. (We never called it 'the belt', or 'the tawse', in the Highlands). Like all my classmates I had a healthy fear of the headmaster's strap. When the ultimate sanction was breached, however, and we had to get the strap, the very last thing any of us did was to tell our parents that we had 'been abused'. Nor, believe me, did we bear any grudge. We knew that we had merited the punishment – we had taken the risk – we did not resent the retribution. I may say that I cannot remember a girl in our class ever getting the strap nor, I must confess, did I ever hear any one of them talking of 'equal rights' in that respect. Probably the girls were just better behaved.

The Battery rock lay just beside the school; it was so called because it was said that Prince Charles Edward Stuart's Jacobites used the rock as a site for a captured battery of guns to bombard the fort from which Fort Augustus took its name. This rock, thickly whin-covered, was strictly out of bounds to us school children, thereby rendering our illicit visits to it the more alluring – to us! This was one transgression, incidentally, for which the penalty was the strap – if you were caught.

We often hear criticism nowadays that there are no policemen any longer walking the beat – that they are all in cars! Our village bobby was always on the beat, on his heavy, black, 'sit-up-and-beg' bicycle. His son, Willie, was in school with us and, inevitably, we called him Willie 'Baton', instead of his rightful surname of Fraser. It was with this Willie and one or two other boys that I very nearly fulfilled my elders' prophecies of an untimely end. As with most boys, water had a great fascination. This included the canal, deep, black, sullen and with a well-merited reputation for accidental drowning fatalities. Large, unwieldy barges lined one bank of the canal, at the top lock. These were used, as necessary, to carry cargoes up or down the canal. These usually empty barges were very strictly out of bounds; despite this prohibition we, of course, played on them. Willie 'Baton' one afternoon decided to try a new fishing rod on the canal; we went with him. The top lock gates were tried first, fruitlessly; Willie decided to try the vantage point of the barges. Great idea! We ran all the way, with alacrity!

Fishing, and hoping, got under way – Willie somehow got his line in a tangle. Would I help him to untangle it? I would! I then proceeded, foolhardily as was my wont, to walk backwards along the narrow rim which bordered the capacious hold of the barge, with below me, quite unregarded, the black canal, while pulling out the line which he unwound from his reel. I must stress that this daft idea was not Willie's – it was mine. Walking nonchalantly backwards along the rim I missed my footing, staggered and swayed, and went down into the canal. I had just sufficient sense (you have leave to doubt this) to dread that I might get wedged under the curving bottom of the barge and so be unable to surface. I therefore tried to kick away from the barge with my feet against it. I sank, and came up spluttering. The entire short span of my six-year-old life did not flash before my eyes; I pushed again with my feet against the barge waterline, and sank again. When I came up this third time I saw the thrust down fishing rod with which Willie was doing his utmost to reach to me. I grabbed for it, felt myself being slowly lifted, then as the weight of my body came full on the rod, as the cushioning of the water ceased, so did the rod joints come apart, and I sank once again. I had been aware of the row of frightened heads peering down at me, including that of David my younger brother who was crying now. I was not aware that a young man, Hughie Grant, had heard the commotion and was pounding up to the barge, holding a boathook, having summed the situation up. When I surfaced, a trifle water-logged now, I was ignominiously gaffed, and hauled, dripping, out. More ignominy was to follow. David held my hand in case he would yet lose me and we trudged the mile or so home, leaving a watery trail behind. I was, by now, putting an explanation through my mind to avoid too severe a punishment. I was, truly, not really conscious that I had, narrowly, just escaped being another statistic on the canal's roll of dishonour. Retribution mercifully, filled my juvenile mind.

Apprehension vanished into utter indignation when my other younger brother, Hugh, who had stayed at home, spotted, out of the kitchen window, my drowned rat appearance and naturally burst out laughing. Years later, he in turn was to rescue me from drowning in the spate-swollen waters of the river Tarff, while on a fishing expedition. I could swim by then but that was a useless attribute in these raging rock-constricted waters on that day. Only his lightning-fast reaction, in thrusting out the walking stick he carried so that I could grab this, in the split second before I was swept away, literally flat out, by the current, saved me. Once again I was hauled out, ignominiously, dripping wet.

More primary school day memories are of juvenile raids on apple trees in the darkness of autumn nights and being one of a coterie of admiring children peering at a gigantic, fresh-caught salmon, which dwarfed us, hanging on the whitewashed gable end of a crofter's cottage, its wide fluked tail trailing on the ground. It had been caught, legitimately, on Loch Ness. It remains the largest salmon, in girth and gleaming length, that I have ever seen. I doubt if salmon of that calibre will be taken out of Loch Ness again.

I also recall a first exciting, clandestine cigarette, at six years old, from a packet bought for 6d (2½ p) and shared with a school friend, Jimmy, as we crouched under the old railway swing bridge (no longer there) which was used to cross the canal above the top lock. I am glad to say that I never took up the habit, not through any virtue, not through any medical advice, for there was none then, but simply because I never liked the taste.

You will see then that I had a happy early childhood in Fort Augustus despite the entirely justified misgivings of my elders.

There followed schooling in Glasgow, shortened by the War, with its particular memories of the banshee notes of air raid sirens and of infinitely ugly gruesomely jagged bits of hot shrapnel, a piece of which I contrived to pick up during a raid. My youthful imagination, always too vivid, conjured up horrifying images of what this could do to human flesh. Cherished holidays punctuated these Glasgow days, always to Fort Augustus, via the 6 am train up the West Highland railway. I went out with my Uncle Bill, a 'god' in my eyes since he let me shoot a rabbit with a 12-bore shotgun which feat left me with an immense joy but also a splitting headache and a sore shoulder. Rabbits, in the War years, were highly esteemed as an addition to the meagre meat ration. Moreover my Aunty Nancy had developed a duodenal ulcer, for which white meat was recommended. Both chicken and rabbit were regarded as white meat but chicken, in the stringencies of those days were very rarely available. There were however sundry ways and means of obtaining rabbit, all of which I learnt.

I saw the Home Guard (called the Local Defence Volunteers at first) formed in Fort Augustus. Both my Uncle Bill and Hugh, my brother, were in this, a trifle solemn and 'Dad's Army' like in their khaki uniforms but committed nevertheless. Weird doings occasionally were attributed to the Home Guard in the Highlands; logically enough, in those days of meat shortage; a red deer stag or a hind sometimes fell to a military rifle. There were no deep freezers then available – the venison secured was shared out among relatives and friends, which included an urgent perishable parcel to Glasgow. I remember too, (now with some amazement and awe), when on holiday from Glasgow, practising with a Home Guard Bren machine-gun with live ammunition, quite unauthorised of course. We thought that we were quite safe from interference, remote on the top of the Crofter's hill, above Loch Unagan. Strangely, later, much later, in my Army years, I was to score a possible maximum points with the Bren light machine-gun, on an authorised battalion target shoot this time; the old Mark 1, a lovely weapon. Of all the tales which went around, derisively and apocryphally, perhaps the best was that which had the entire Lochaber section called out, one autumn night, when, it is avowed, the OIC mistook a huge skein of airborne, migrating geese for an invading German airborne invasion. Did we believe it? I really don't remember, but I know we certainly enjoyed the thought.

Inevitably, while still in Glasgow, I was sucked into the war machine before that particular madness, the Second World War, was over. Aberdeen, Yorkshire, Staffordshire, all followed. In Staffordshire there occurred a first which I cannot credit to Fort Augustus. This was my meeting with Margaret. Margaret has always maintained that she did not whistle at me that day on her own behalf; it was on the totally unselfish behalf of a friend who could not whistle. It was undoubtedly my good luck that Margaret could whistle; some years later, after I came home from abroad, we were married. To that Staffordshire whistle I owe a very great deal.

Overseas, India and Burma followed and then after the War, twenty-one happy years at Fort Augustus with our two sons growing up fast.

I attempted to learn red deer management the hard way. My predecessor, old John MacDonald (better known locally as Johnny Kytra) had memories right back to the recruiting drive for the Lovat Scouts, for the Boer War. He had been a piper in the 1914–18 War; he had been too old for the Second World War. His had been a different era. 'Selection?' he said to me, quirking a bushy eyebrow while he lit his squat pipe, clouds of pipe smoke, aromatic, wreathing his head, 'we did not bother selecting which red deer to shoot; it was down with them, whenever you had a chance!' Yet Johnny was not a cruel or callous man; he was a veteran of a harder era; his had always been a simple, down-to-earth, struggle for survival. There was no leisure to reflect upon the imponderables of life. Johnny's dictum would have been 'only the well-off can afford to be thinkers.' Johnny was never 'well-off' but he enjoyed life nevertheless, at times with gusto. Every Saturday night would see him cycling the four long miles home, from a session in the Lovat Arms, an ancient, camouflage, gas cape (relic of World War II) billowing out behind him. I grew to value his down-to-earth knowledge, and his friendship, which lasted until he died, at ninety years old.

I learned to play badminton and I also tried shinty and succeeded in getting some teeth knocked (literally knocked, for a shinty stick can be quite uncompromising) into the roof of my mouth. They were artificial teeth I should add; I had already lost one set in the Indian ocean, now I had 'lost' one replacement set on the shinty field. Nevertheless I enjoyed shinty, once my mouth healed.

All these associations are not readily forgotten or ignored – do you wonder that I was determined to retire, back home, to Fort Augustus? This then, quite unsuspectingly, was my apprenticeship for Torridon.

Out of the blue, came the need to decide whether to go to Torridon or not. I believe that I made the correct decision. Forgetting all the superlatives, all the superb scenic attributes, I found that Torridon is basically harsh and unrelenting country, tolerating few mistakes. This, of course, is also intrinsic in the allure of Torridon, as it is of the Cuillins of Skye, the Glencoe hills and An Teallach, they are all wild, independent areas, appealing to independent characters.

To enjoy, to fully appreciate, the perpetual challenge which is Torridon one should be young and, above all, fit; young, perhaps, in spirit more than in actual years, and definitely not young in rashness of outlook. I admit to being forty-two years of age when I went to Torridon, well past the first exhilaration of youth. Nevertheless I was physically very fit and had had very many years of experience of hills, admittedly much less demanding hills than those of Torridon.

Torridon was never easy, nor will it ever be easy, if one has a feeling for the hills and the required respect for them and all that pertains to them. It is not fortuitous that wild areas of land in the Highlands, of which Torridon is only one example, have stayed wild. They have defied the acquisitiveness, the inventiveness and the greed of mankind yet when one is determined to stay with, and to accept these 'disadvantages', when one lacks the 'ambition' to seek the crowded cities of the world, and their 'opportunities', there is no better place to live in than the Highlands.

Until I became sixty or so I was happy that I was able to cope with all that Torridon could throw at me even in its most brutal moods. After that, slowly but relentlessly, Torridon began to grind me down – a successful cartilage operation in my right knee then, (as I had been warned might happen) osteo-arthritis crept in to both knees. I accepted, not altogether gracefully, the restrictions on my former activities, notably on my annual treks to eagle eyries, and my participation in mountain rescues. Then came further limitation to my lifelong enjoyment of solitude, of high, remote and dangerous places where solitary self-reliance was the key. While dragging in three deer carcases, on an intensely cold and frosty January morning, the rest of the world still at its breakfast, I sustained what was later said by the doctor to be a slight heart attack. Thereafter, I have to admit, I struggled along as best as I could until retirement.

I shall definitely never be able, nor to be willing, to forget Torridon – it is magnificent country and it was never designed to be subdued or subservient to mankind's commercial interests. Torridon tolerates rather than welcomes us – we need its wild breathing spaces, its wild places, its lochs and high ridges, as an antidote to our over-urbanised, over-commercialised, over-trivialised, dreadfully overcrowded world.

To me Torridon has many memories: its varied moods, its tranquility, its aloofness, sunsets, sunrises, rainbows, salt-laden gales, drenching rains, and, occasionally, serene snow-scapes – its neighbourliness in times of need, when help was not measured by a payment expected; its relative absence of sneaky nastiness (and our nearest policeman was twenty-three miles away); its diversity of wildlife on sea and on mountain. Life is not idyllic there, it has its own problems, but it is freer than in most places.

There remain many cameos. That first airy traverse of Liathach and its Pinnacle ridge, 'when all the world was young', and when all of it seemed spread below our feet. That night out, in the coire of the Wolf's hole on Beinn Alligin; a rather different night out when at a New Year's dance (which took place in what was normally a cafe) a youngster passed out, quietly and without fuss, and was 'laid out', also quietly and without fuss, on top of the long chest-type deep freeze at one end of the dance floor, to recover. Most appropriate it seemed at the time. Rescues on the mountains; the comradeship of a team doing a dangerous job, voluntarily and the occasional utter sadness in being able only to bring a body home, off the hill. The eagle, which, unasked and unexpected, flew in from the sea; that utterly fearless otter, in the snowy ditch, on a March morning; frogs, uncountable, spawning in that same ditch and Joann Robertson, six years old, asking her Uncle Bob to catch her a frog, then, on being given one, rejecting it, stating, indignantly, that she wanted 'a double decker'. The plaintive piping of the ringed plover as it was borne, fast, away in the feet of the peregrine falcon; the ringed plover nest 'protected' by the old tea pot. The red-throated divers which flew inland from the sea each breeding season, gabbling harshly and discordantly as they winged their way curiously hump-backed, to a breeding loch. Pine martens, lithe, glossy-furred, incredibly agile, as one tried to capture their vitality on film. A three-cornered golden eagle aerial fight when Tom Wallace and I watched a 'strange' eagle being escorted 'over the frontier' by the resident pair. Twice the interloper was forced to the rocks, grounded, by one of the territory eagles, while the other kept station high above, before it was at length deemed driven far enough away. The extremely rare birth of the red deer twins to my hind, Beauty, a character in her own right; the nasty taste in the mouth left by the needless death of my four-year-old Royal stag. Innumerable hill walkers at the end of a day's

walk, invariably grateful, invariably expressing enjoyment, even as
the rain dripped, oozing in liquid droplets, from them. How could
one forget these little gems of life?

Of Margaret – well, simply, that, without her help, the twenty-one
years in Torridon would have been quite impossible. She made many
friends and throughout the years she has retained that inner spirit
which makes her ageless. Margaret is a good baker; many an
incipient bureaucratic crisis averted by our joint managerial plan – a
plate of warm scones on the tea table and a bottle of MacAllan in the
sideboard. She has, too, never lost that irrepressible, impish sense of
humour which bubbles to the surface at the oddest moments. You
know, *I do not know*, to this day, whether she was *really* whistling on
that far away day, for her non-whistling friend or for herelf. With the
end result however I am content!

A final word as apposite to Torridon as it is to the Highlands in
general. I quote, from memory, the words in which A. M. Youngson
referred to the Highland area 'It has been the unique excellence of
the Highlands that they are grand, and different. They did not submit
to ordinary kinds of human exploitation. For ages they preserved a
culture of their own and an independent way of life. For longer still
they have remained, like Northern Norway and Sweden, like parts of
Arizona and New Mexico, remote, unexploitable, away from the
great centres of industry and of population, little affected by man and
modern technology. It may be that it is this, not any potential for
being re-fashioned in the image of Lanarkshire or of Midlothian, that
gives the Highlands their highest value in the modern world'.

I truly believe that this is indeed so!

Index